Smart Buildings

Jim Sinopoli

Visit www.smart-buildings.com for more information and resources on smart buildings.

ISBN 0-9786144-0-2

Library of Congress Control Number: 2006904723

Foreword

After being involved with hundreds of building construction projects, I find it ironic that technology has enabled globalization, where places, people, and systems around the world have become connected and interdependent, yet many buildings are constructed with separate and independent building technology systems. How is it that we can have instant worldwide communications of voice, data, and video, but have difficulties deploying inter-networking within a building?

It seems that any hesitancy one has in designing or operating a smart building has very little to do with the state of technology, which is readily available and rapidly evolving. Instead, it has everything to do with legacy facility design and management processes which tend to isolate each technology system into a separate silo.

Even though smart buildings have been constructed worldwide, their full potential is still waiting to be fully recognized by those involved in the design and operation of our buildings. A large part of this is related to a lack of understanding of all the technology systems involved, as well as the lack of clear and succinct information on how to truly integrate and converge the systems.

The intent of this book is to bridge that chasm and provide simple, straight forward information on smart buildings for architects, engineers, facility managers, developers, contractors, and design consultants. *Smart Buildings* is a practical guide and resource for designers and facility managers, covering the basic design foundations, technology systems, and management systems encompassed within a smart building.

The basis for smart buildings are, quite simply, a handful of technical standards that already dominate the marketplace – Ethernet, TCP/IP protocols, SQL databases, standard fiber optic and unshielded twisted pair cables, and the Internet. I believe that if you adhere to these underpinnings, you will be well on your way to a smart building.

Unlike other resources, *Smart Buildings* is organized to provide an overview of each of the technology systems in a building, and to indicate where each of these systems is in their migration to and utilization of the standard underpinnings of a smart building. Specialists of certain technologies may find the coverage of certain technology systems elementary, but can gain knowledge of other technology systems in the building that they may be unfamiliar with. A comprehensive knowledge and grasp of all the building's technology systems is important to the deployment of a smart building.

Technology always has and always will influence the buildings we build. The convergence of communications networks within our buildings is a significant technological and building transformation and it provides some truth to the old adage that the whole is greater than the sum of the parts.

Table of Contents

Table of Figures

What is a Smart Building?

Over the last two decades, developers, property managers, engineers, architects, facility managers, building owners, and system design professionals have all had a slightly different perspective on what makes up a "smart" or "intelligent" building. Unfortunately, these varied perspectives have spawned many definitions of what constitutes a "smart building."

Some definitions are esoteric with wildly futuristic and meaningless explanations like "a smart building is a building that improves productivity and increases facility attractiveness." Some definitions solely focused on telecommunications, such as "a smart building is served with fiber optic cable and provides high speed Internet access." Others simply refer to advanced building automation systems. To some degree, each of these definitions is correct, yet they are also incomplete.

The essence of the smart building is comprised of advanced and integrated systems for building automation, life safety and telecommunications systems. A typical building has multiple technology systems, possibly upward of 14 or 15. The traditional way to design and construct such a building is, to a large extent, to design, install, and operate each system separately (Figure 1).

Multiple Proprietary Building Systems

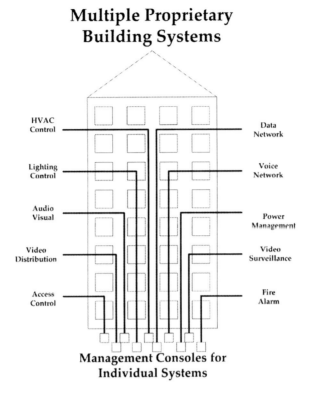

HVAC Control

Lighting Control

Audio Visual

Video Distribution

Access Control

Data Network

Voice Network

Power Management

Video Surveillance

Fire Alarm

Management Consoles for Individual Systems

Figure 1 Multiple Proprietary Building Systems

The systems however, share several common features: they need a communication network between system components; they need cabling, cable pathways, and equipment rooms; and they have system databases and some type of communications rules or protocols between devices. Yet despite these commonalities, we plan, deploy, and run the systems separately.

Essentially, smart buildings are integrating these systems from a physical and logical perspective, the result being more function-

functionality and lower costs for those developing, occupying, and managing buildings (Figure 2).

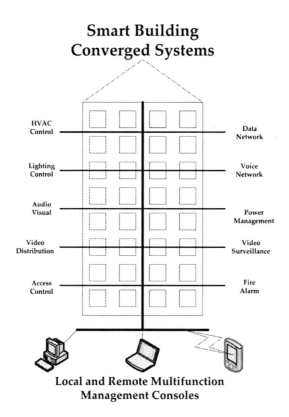

Figure 2 Smart Building Converged Systems

The reasons to build a smart building are twofold. The first is to save money. Savings are related to both the initial construction cost and the overall operations cost of a facility. The second reason is that by integrating the systems the building manager and tenants can do things that simply can not be done with separate systems. Integration means the systems communicate and share data, provide more functionality, and allow information from one technology system to affect the actions of other systems. For example, if a smoke detector alarm is activated, the access control system changes to emergency mode; the heating, ventilation, and air conditioning systems adjust; the video surveillance camera changes so the effected area can be monitored; and so forth. Smart buildings leverage IT infrastructure benefits by taking advantage of existing and emerging mainstream technology.

While the idea of integrating telecommunications, life safety, and building automation systems may have been ahead of its time and ahead of the underlying networking technology in the 1990s, that is the case no longer. Many buildings around the world have such systems, many manufacturers have materials and products available, and technical trends are improving and accelerating the integration of these systems.

For building developers and owners, smart buildings add another level of value to a building. For property and facility managers, smart buildings provide more effective subsystems and more efficient management options, such as the consolidation of system management. For building architects, engineers, and construction contractors, it means combining portions of the design and construction, with resulting savings in project management and commissioning time. It is clear that if one is planning, designing, constructing, managing, owning, leasing, or maintaining a facility, that a smart building approach needs to be an integral part of the programming and deployment.

The Economics of Smart Buildings

Buildings have long life cycles, typically between 25 and 40 years. The life cycle cost of a building includes the initial costs of the facility (concept, design, financing, and construction), as well as the long term operational cost of the building (Figure 3).

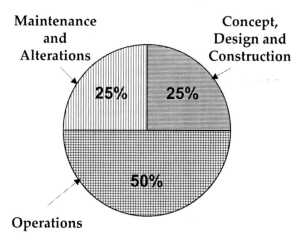

TYPICAL LIFE CYCLE COST OF A BUILDING

Maintenance and Alterations — 25%

Concept, Design and Construction — 25%

Operations — 50%

Figure 3 Life Cycle Building Cost

Smart buildings can reduce both the construction cost of the technology systems, as well as the operational cost of the building. The cost savings from the smart building approach results in added value to the building as evidenced by lower capital (CAPEX) and operational (OPEX) expenses (Figure 4).

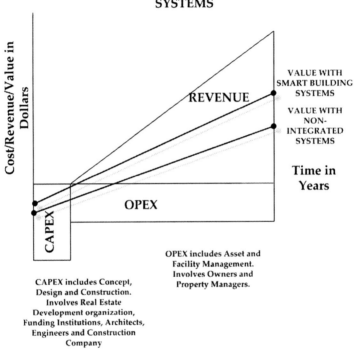

DISTRIBUTION OF A BUILDING'S COST OVER ITS LIFE AND THE ADDED VALUE OF SMART BUILDING SYSTEMS

Figure 4 Added Value of Smart Buildings

Part of any building's construction costs includes the systems that are at the heart of a smart building, namely telecommunications, building automation, and life safety systems. There is a construction cost for these systems whether the systems are installed separately or are built as integrated smart building systems. The construction cost savings of integrated systems is primarily attributed to the efficiencies in cabling, cable pathways, labor, project management, and equipment.

When the technology systems are installed separately, the project has many more contractors or subcontractors involved. Each system's subcontractor is installing a system requiring cabling, cable pathways, equipment space, power, air conditioning, servers and management consoles. Each mobilizes a workforce. Each contractor must be managed, monitored, and coordinated by the general contractor. A great deal of overhead is created this way.

The smart building consolidates the cabling infrastructure and the number of contractors involved in the construction. In an integrated system, the infrastructure cost, consisting of both cable and cable pathways, has a considerably lower cost for labor. Project management and engineering costs are typically reduced as well. The net result is that the smart building approach can be considerably less costly than installing systems separately. Additional installation costs savings can be realized from the following resources as well:

- The consolidation of servers for the systems which results in less hardware, less space and reductions in

software licenses
- The required training of personnel on standardized system management tools and platforms is lessened
- The time to configure systems due to the use of shared standardized databases is lowered
- Prudent use is made of wireless devices, where cabling is cost prohibitive
- Power Over Ethernet (PoE) is utilized rather than local power for devices

The largest cost difference between integrated and separate systems is the long term operational cost of the systems and the facility. The operational cost savings are related to several factors:

- The standardized infrastructure allows for easier change implementations during the operational life of the facility for building automation system controls and devices , telephone systems, data networking, lighting, and other telecommunications and building automation systems
- Increased building efficiency which results in energy savings
- Systems are highly effective, for example, allowing the fire alarm, video surveillance system, access control system, HVAC system, the lighting control system, and the elevator system to all communicate during an emergency evacuation
- Standardized management tools continue to reduce training costs
- Improved information management on the systems
- Improved overall staff productivity
- The ability to competitively procure systems with open ar-

8

chitectures and a generic structured cabling system

- The consolidation of system management among many separated facilities through the use of the Internet or a private network
- Easy integration with additional business systems, such as Human Resources and Purchasing due to standardized databases

For a building lasting 40 years, operational cost may amount to 50% of the total cost of the building. It's equal to the total cost of construction, financing, and renovating the building. Thus a relatively modest savings in annual operational costs will garner significant savings over the building's life cycle.

Other organizations, both public and private, have weighed in on the cost savings of integrated smart building systems:

- A study conducted at the National Institute of Standards and Technology, posed the question, "...does it indicate that investment in CBS (cybernetic building systems) products and services by individual owners and operators will be cost effective? The answer to that question is most certainly yes." They concluded that "For every dollar invested in 2003 approximately $4.50 is returned (i.e., a savings-to-investment ration of 4.5). This equates to an adjusted internal rate of return of approximately 20% per year..."[1] The study conservatively estimated that inte-

---

[1] Chapman, R.E., February 2001, *How Interoperability Saves Money*, ASHRAE Journal,

grated systems have annual energy cost savings of $0.16 per sq. ft., annual maintenance savings of $.15 per sq. ft., annual savings for repair and replacement of $0.05 per sq. ft., and annual savings related to "occupant productivity" of $0.39 per sq. ft.

- The Continental Automated Building Association analyzed life cycle costs[2] and concluded that:
  - First costs for integrated systems (including management hardware and software, network upgrades, web services, and reductions in devices) were 56% less than non-integrated systems
  - Annual costs for changes, alterations and upgrades, after a system's warranty period (including service contracts, additions and remodeling, software upgrades and reserves for systems replacement) for integrated systems were 32% less than non-integrated systems
  - Annual operating and maintenance costs (such as staff, training, IT support, and management reporting) for an integrated system are 82% less than a non-integrated system.
  - Integrated systems saved 10% of utility cost (including integrated lighting and HVAC, improved load factor, coordinated supply and demand strategies)

---

[2] Continental Automated Building Association, *Life Cycle Costing of Automation Controls for Intelligent and Integrated Facilities*, A White Paper for Task Force 3 of the Intelligent and Integrated Building Council, April 2004

as compared to energy costs for non-integrated systems.

- o The Net Present Value of the life cycle costs of an integrated system (10 years with a discount rate of 9%) was 24% less than a non-integrated system.

- Systimax, a Commscope Company that manufacturers and distributes the physical layer (cable, cable pathways, connectors, etc.) for smart building systems prepared a cost model[3] primarily related to initial cabling and infrastructure costs for technology systems, and the ongoing costs related to the "churn-rate" of moving, adding and changing services. The model was based on a typical 100,000 SF 5-story commercial office building, using the same cable for all technology systems and a common pathway for all horizontal low and high voltage services. Systimax concluded:
  - o The cable and cable pathway installation costs for integrated systems were 16% less than non-integrated systems and needed 44% fewer labor hours
  - o The cable and cable pathway costs for addressing moves, adds, and changes over a five year period for integrated systems were 39% less than non-integrated systems

---

[3] Systimax, *Cost Reducing Construction Techniques for New and Renovated Buildings/Cost Models*, White Paper Issue 2 March 2004

SUMMARY OF COST ADVANTAGES OF

SMART BUILDING TECHNOLOGY SYSTEMS

INTITAL COST

Management hardware and software, network upgrades, web services and reductions in devices	56% Less
Cable and cable pathway installation cost	16% Less

OPERATIONAL COSTS

Service contracts, additions and remodeling, software upgrades and reserves for systems replacement	32% Less
Annual savings for maintenance, repair and replacement	$0.20 per sq. ft
Staff training, increased staff efficiency, IT support, and management reporting	82% Less
Integrated lighting and HVAC, improved load factor, coordinated supply and demand strategies	10% Less
Utility cost	$0.16 per sq. ft
Cable and cable pathway cost for addressing moves, adds and changes	39% Less

LIFE CYCLE COSTS

Net Present Value of the life cycle costs of an integrated system	24% Less

Figure 5 Summary of Smart Building Cost Advantages

The Foundations of a Smart Building

Smart buildings are built on open and standard communications networks. This allows technology systems for: (a) different applications to communicate with each other, (b) efficiencies and cost savings to be garnered in materials, labor, and equipment, and (c) interoperable systems to be procured from different manufacturers. The primary technical underpinnings of smart building technology are the use of:

- Structured Cable Infrastructure
- IP, TCP/IP, and Ethernet Protocols
- Interoperable System Databases

Structured Cabling Systems

The authoritative standard for cable infrastructure for telecommunications systems within buildings is the ANSI/TIA/EIA-568 A and B Commercial Building Telecommunications Cabling Standard, published by the American National Standards Institute (ANSI), the Telecommunications

Industry Association (TIA), and the Electronic Industry Association (EIA).

Recently these standards organizations also published ANSI/TIA/EIA-862 Building Automation Systems Cabling Standard for Commercial Buildings, which is essentially documentation of similar standards for building automation systems. These two standards are applicable to all of the smart building technology systems.

Other applicable standards include:

- TIA/EIA 569-Commercial Building Standard for Tele-communications Pathways and Spaces

- TIA/EIA 606-Administration Standard for Commercial Telecommunications Infrastructure

Both the ANSI/TIA/EIA 568 and 862 standards allow for an open cable infrastructure that is independent of any products and vendors. The primary basis for both standards is the use of unshielded twisted pair (UTP) copper and fiber optic cable. Both standards have similar design guidelines and parameters for reliability, capacity, and compatibility.

However, there are two minor differences in the BAS standard:

- A "horizontal connection point (HCP)" which is similar to Telecom's "consolidation points" or zone cabling
- A "coverage area" as opposed to Telecom's "work area"

Despite the minor differences, these two standards allow for a single cabling system for a building. Until the recent standard on BAS cabling, building automation systems were cabled separately using different cable types and cable pathways. In addition, these systems have traditionally used hardwired connections of cables from the BAS equipment to the devices, unlike the ubiquitous twisted pair RJ-45 connections of the telecommunications network world. All of this is currently evolving to increased use of standard cable infrastructure based on unshielded twisted pair copper and fiber optic cables.

It should be noted that twisted pair copper and fiber optic cables are used extensively in these networks, but not all end-devices warrant the use of such cables. For example, twisted pair cables may be used to connect a personal computer to a network, but would not be used to connect a mouse or key-board to the personal computer. Similar situations exist with other end-devices in other smart building systems.

Twisted Pair Copper Cable

The core of a twisted pair copper cable (Figure 6) is two insu-lated copper cables twisted together into a "pair." Four pairs are jacketed together for a standard 4-pair copper cable. The wires are relatively thin (between 22 and 24 gauge). The cables are twisted into a pair to reduce crosstalk ("coupling" of the pairs or interference from an adjoining cable) as well as inter-ference from electrical and mechanical sources. Each pair has a different number of twists relative to the other pairs in the cable to further reduce crosstalk. This construction is usually referred to as unshielded twisted pair.

The twisted pairs can also be "shielded," creating shielded twisted pair cable. While the use of shielded twisted pair is popular in some countries and has some technical advantages, it is more labor intensive during installation because of the grounding required, and is typically not used in a structured cable infrastructure.

Figure 6 Unshielded Twisted Pair Cable

Unshielded twisted pair cables are relatively inexpensive and there are a large number of technicians qualified to install the cable. The cable standards guarantee performance of the cable over the distance of 90 meters or 295 feet. Categories of unshielded twisted pair are based on the bandwidth, or information carrying capacity, of the cable.

The most recent categories of unshielded twisted pair are Category 5e and Category 6. Category 5e cable is specified up to a bandwidth of 100 MHz (Hz is a unit of frequency equal to one cycle per second). The Category 6 standards set requirements up to 250 MHz. These standards not only apply to the cable but also to all connector and cable termination devices, such as the cable jacks, the patch cords and the patch panel. Many manufacturers test and/or manufacture cable beyond the standards to differentiate their product as going beyond the standards.

Fiber Optic Cable

Fiber optic cables use strands of glass to propagate light. The light pulses transport communication signals between devices. At the center of the fiber optic strand is a small inner core which carries the propagated light. Surrounding the core is the outer cladding. Both the core and the cladding are glass, but have different "refractive indexes," which essentially means that light travels at different speeds through the materials. The result is that light pulses produced from lasers or LEDs at one end of a fiber optic cable are sent through the fiber optic core and are reflected back to the core when the light hits the fiber optic cladding, thus keeping the light within the center core.

Fiber optic cables with small inner cores (cores of 10 microns or less) have only one path for the light and are referred to as single mode fiber. Fiber optic cables with slightly larger cores (50 and 62.5 microns) have multiple paths for the light and are referred to as multi-mode fiber. The cladding of both types of fiber is 125 microns. For comparison, human hair is generally in the range of 40 to 120 microns (a micron is one millionth of a meter, or about .00004 of an inch).

As the light pulse transverses the cable, it loses power. This loss, called attenuation, is measured in decibels. The attenuation of fiber optic cables is dependent on the wavelength. Multi-mode fiber optic cables operating at low wavelengths may have attenuation less than 3.5 decibels per kilometer and less than 1.5 decibels per kilometer at higher wavelengths.

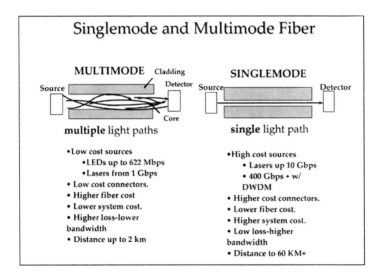

Figure 7 Fiber Optic Cable

Single mode fiber optic cables have superior performance with maximum attenuation of 0.5 to 1 decibel per kilometer, and can be used anywhere in a network, but is generally used for longer distances. Multi-mode fiber optic cables are used for shorter distances, less than a mile, and are generally used within buildings.

Beyond the high bandwidth capacity of fiber optic cables, major benefits include:

- Immunity from electro-mechanical interference
- Immunity from radio frequency interference
- Use over longer distances

Coaxial Cable

For many years coaxial cable was the cable of choice, with extensive use in video distribution systems, early Ethernet network installations and connectivity to many large main-frame computer systems. Coaxial cable is essentially a waveguide, transmitting radio and television frequencies down the cable, and is immune to electro-mechanical interference. It continues to be used for video transmission. However, with technical advances in IP video, and the use of "baluns" (which allow video signals to be transmitted over unshielded twisted pair), the use of coaxial cable is decreasing. Its use within a smart building is generally minimized.

Wireless

Wireless connectivity is just a substitute for cabled connectivity. Wireless does not, and technically can not provide the bandwidth of a physical cable connection. However, wireless can provide mobility. It is also a very viable option for connectivity in older buildings where pathways for cable may not be available. The wireless technologies probably most useful for smart buildings technology systems include Wi-Fi and an emerging technology, Zigbee.

Wi-Fi

Wi-Fi basically replaces a cabled Ethernet connection with a wireless device. Current "Wi-Fi" (Wireless Fidelity) systems operate in two unlicensed radio frequencies, 2.4 GHz and 5 GHz. The Institute for Electrical and Electronics Engineers (IEEE) has set three standards for Ethernet communications via these frequencies which are commonly referred to as IEEE 802.11a, operating in the 5GHz frequency, and IEEE 802.11b and 802.11g, which operate in the 2.4 GHz frequency. These standards can provide "optimal" throughput of 11 Mbps (a measure of bandwidth, megabits per second) and 54 Mbps. The user's distance from the antenna, the uses of the same unlicensed frequencies by other devices, and the obstacles within and the structure of buildings which could interfere with the radio signals all affect the communications bandwidth received from the Wi-Fi antenna.

Typical coverage areas indoors for omni-directional Wi-Fi antenna are 100 to 300 feet. Each "wireless access point" (WAP) or gateway can generally serve 10-20 users, depending on the users' applications. This technology and wireless "hotspots" are now common in public buildings, airports, businesses, hotels, restaurants, and homes. The marketplace for Wi-Fi equipment is now moving toward "wire-line-class" security, high performance, reliability, and enterprise-scale manageability of systems.

Wi-Fi Mesh Networking

Instead of using cable to connect all wireless antennas, a wireless mesh network has all nodes or WAPs interconnected wirelessly. Typically, these "hotspots" are created using omni-directional Wi-Fi IEEE 802.11b/g antennas. Directional antennas using Wi-Fi IEEE 802.11a are then used to connect the hotspots or nodes and create the mesh. The 802.11a standard is used in the mesh backbone because of its performance (54 Mbps) as well as its different and non-interfering radio frequency (IEEE 802.11a uses 5 GHz, while 802.11b/g antennas use 2.4 GHz).

The nodes on the mesh network automatically learn about one another and self-configure network traffic through numerous network path configurations. The result is extended coverage of a Wi-Fi network with the ability to route traffic around congestion and around obstacles and interference. Mesh networks provide redundancy and "robustness" because of the capability to balance network traffic. Wireless mesh networks can create some latency which can effect applications such as VoIP. They can also be noisy, possibly creating transmission errors, resulting in re-transmissions and reductions of available bandwidth. While current implementations are proprietary, a standard for Wi-Fi Mesh networks, called 802.11s, is being developed, and is expected to be finalized and released in 2007.

Zigbee

Zigbee is an emerging wireless technology standard (IEEE 802.15.4) which provides for low data rate networks. It uses unlicensed frequencies (900 MHz and 2.4 GHz) which are also available for cordless phones, Wi-Fi, and other devices. The standard is aimed to address residential, building, and industrial control devices. It is specifically useful for sensors and control devices of building automation systems within a smart building where very small amounts of information or data are being transmitted. This includes on/off switches, open/closed devices, thermostats and motor controls. The maximum speed of Zigbee devices varies up to 192-250 Kbps (a measure of bandwidth, kilobits per second). The maximum distance varies between 20 and 50 meters.

Zigbee has several advances:
- Low power usage as the devices only require two AAA batteries
- Wide support from more than 100 companies support-

- ing the standard (companies such as Motorola, Honeywell, Samsung, Mitsubishi, and others)
- Mesh technology which allows Zibee, like Wi-Fi, to be configured in several topologies, including a mesh topology allowing mutliple transmission paths between the device and the receiptient
- System scalablity where thousands of Zigbee devices can deployed within a building

Ethernet and IP Protocols

Many years ago, data processing was done solely on mainframe computers or other centralized computers. These systems had proprietary communications between the user's "dumb" terminal and the mainframe equipment, both of which were manufactured by the same company. For example, one could not take a user's terminal manufactured by IBM Corporation and use it on a system made by Digital Equipment Corporation.

Today's data networks use personal computers and devices from multiple manufacturers. Different personal computers can be networked, not only because they have processing power and storage that the "dumb" terminals did not have, but more importantly, because they communicate through a common, open communications protocol. The technology systems in a smart building have adopted and are migrating to the common network protocols used in data networks.

At the forefront of the evolution to open network standards is the International Standards Organization's (ISO) development of the Open System Interconnection (OSI) model (Figure 8). The OSI model presents seven layers of network architecture (the flow of information within an open communications network), with each layer defined for a different portion of the communications link across the network.

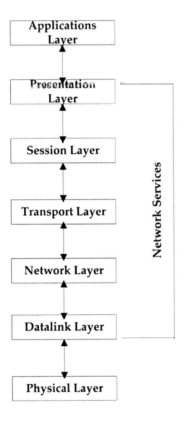

Figure 8 OSI Model

A network user creates and initiates the transmission of data at the top layer, the application layer. The data from the sending device moves from the highest layer to the lowest layer, to communicate the data to another network device. At the receiving device the data travels from the lowest layer to the highest layer to complete the communications. When the data is initially sent, each layer takes the data of the preceding layers and adds its own information or header to the data. Basically, each layer puts its own "envelope" around the preceding "envelope." On the receiving side, each layer removes its information or "envelope" from the data packet.

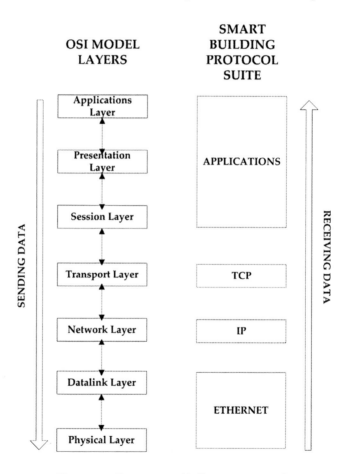

Figure 9 Smart Building Protocols

For smart buildings, the most important layers of the model are the lower layers. The lower layers are as follows:

- Physical Layer
  This layer defines the electrical (in the case of fiber optic cables, light) communications across a network link or channel. This layer guarantees that bits of data transmitted by a device on the network are accurately received by another device on the network. The physical layer initiates, maintains, and terminates the physical connection between network devices. It defines the mechanical and the electrical characteristics of the physical interface, including connectors, network interface cards, and voltage and transmission distances.

  The network protocol RS-232, previously heavily utilized in telecommunications and data networks, is defined solely by the physical layer. Other network protocols, such as Ethernet (Figure 9), are defined by the physical layer as well as some or the entire next layer in the stack, the data link layer.

- Data Link Layer
  The data link layer takes the data bits and "frames," and creates packets of the data to guarantee reliable transmission of the data. The data link layer adds source and destination addresses to the data stream as well as information to detect and control transmission errors. The data link layer has two sub-layers. One is the Logical Link Control (LLC) sub layer, which essentially maintains the communications link between two devices on the network. The other is the Media Access

Control (MAC) sub layer, which manages the transmission of data between two devices. The network card on a personal computer has a MAC address, essentially a unique address for every device on a local area network.

The details of the data link layer can be specified differently and are reflected in various network types (Ethernet, Token Ring, etc.). Each network type has its own method of addressing, error detection, control of network flow, and so forth.

Figure 10 Typical Ethernet Connection

- Network Layer
  The network layer routes data packets through the network. It deals with network addressing and determines the best path to send a packet from one network device to another. The Internet Protocol (IP) is the best example of a network layer implementation.

27

- Transport Layer
  The transport layer is responsible for reliable transport of the data. At times, it may break upper layer data packets into smaller packets and then sequence their transmission. The Transport Common Protocol (TCP), one of the major transport protocols, is typically used with the best known network layer protocol, IP, and is referred to as TCP/IP.

The general technology trend is for everything to become a part of a node on a cabled or wireless Ethernet network (Figure 10) and have an Internet Protocol (IP) address. This is the digital convergence trend we have already seen and will continue to witness for years to come.

In the data telecommunications arena, IP and Ethernet have been standards for years. Voice communications, traditionally modeled after legacy monolithic mainframe computers, are quickly moving to the data telecommunications world, via Voice-Over-IP (VoIP). Video is also moving to digital communication protocols. Building automation systems have their industry protocols (BACnet, LonTalk, and others), but they too are moving to convert or interface the protocols to the universal, dominant IP protocol. Life safety systems still lag behind though video surveillance systems have moved to IP, and access control is moving in that direction as well. Even legacy systems with proprietary standards can have their protocols converted or translated to the standard IP protocols.

Interoperable Smart Building System Databases

Each technology system in a smart building has some sort of data or database associated with its operation. This data may also be needed by another technology system, may be partially duplicated in another technology or management system, or may be needed by a business administration system. Major database standards allow for the access to or transfer of database information within smart building systems.

Structured Query Language

Structured Query Language (SQL) is a standard (ISO and ANSI) that defines rules for the definition, structure, operation, manipulation, and management of relational databases. The first SQL standard was adopted in 1986 and there have been several additions, expansions, and modifications of the standard since then. SQL is a vital and integral part of open network architecture necessary for smart buildings because it allows for databases from different manufacturers to interoperate and exchange data (Figure 11).

IBM is recognized for developing the initial SQL format and rules that were later used to draft the SQL standard. Other major manufacturers have adopted the standard and vend products compliant with SQL, including Microsoft, Sun, Oracle, and others. Many of these manufacturers have proprietary add-ons or extensions of the SQL standard. SQL can run on a variety of hardware (PCs, servers, mainframes), a variety of networks (local, wide, enterprise), and a variety of operating systems (MS Windows, UNIX, Linux, Mac).

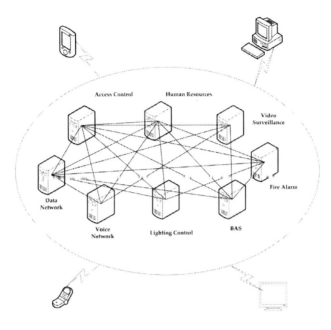

Figure 11 Database Sharing

SQL uses a row and column structure much like the spread-sheet applications used by many personal computer users. The initial purpose of SQL was to make it easy to query databases, but SQL has evolved to a full complement of programming, security, and management tools. Users can query data and programmers can program with simple sentences. Because the programming and user interface is simple and intuitive, data is more accessible and usable.

Open Database Connectivity

Other software applications typically work with SQL databases to provide additional functionality, connectivity, interoperability, and access. One of these is the Open Database Connectivity (ODBC) interface.

ODBC allows ODBC-compliant software applications to access ODCB-compliant databases through a middle layer of software (known as database drivers) that resides between the application and the database (Figure 12). Developers of the application do not need to know the specific database the application will use. A software application can access several databases with multiple drivers.

ODBC is an SQL-based interface developed jointly by Microsoft and an industry working group called the SQL-Access Group, but it is supported by other major manufacturers of SQL databases as well. This interoperability is important for open, integrated smart building systems. For example, data in a spreadsheet can easily be imported or exported to an ODBC compliant database.

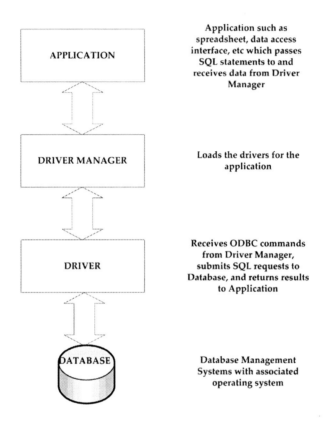

Figure 12 ODBC Operation

Figure 13 Comparison of Systems and Standards

SMART BUILDING SYSTEMS	STANDARD STRUCTURED CABLE	ETHERNET AND IP PROTOCOLS	STANDARD DATABASE
Data Networks	●●●	●●●	●●●
Voice Networks	●●●	●●●	●●●
Audio Visual Systems	●●	●●	●
Video Distribution	●●	●●	●●
Fire Alarm Systems	●	●	
Video Surveillance Systems	●●●	●●●	●●●
Access Control System	●●	●●	●●●
HVAC System	●	●	●●
Electric Power Management System	●●	●●	●●
Lighting Control System	●●	●●	●

●●●   Extensive Use of Standards

●●   Use of Standards

●   Some Use of Standards

33

Data Networks

Data networks are important for several reasons but they are particularly important for smart buildings. This is because the basic infrastructure of data networks (standard cable infrastructure and network protocols) are proliferating and are being adopted by other building systems. These basic data networking technologies are the technical core of a smart building.

Data networks are used to share resources and exchange information between network users and other networks. Decades ago data networks consisted of mainframes, mini computers, and both proprietary networking infrastructure and communications protocols. Today, most networks consist primarily of switches, servers, industry standard operating systems, network and client software applications, peripheral devices, and, of course, user devices.

Users typically use desktop or notebook forms of personal computers or Personal Digital Assistants (PDAs) to connect to the network. The network connectivity of the user to the network is either standardized cable or a wireless access point. Remote connectivity to the network is typically provided through a telecommunications service provider via a cable modem, DSL, T-1 lines, or higher capacity telecommunications services.

A Local Area Network (LAN) (Figure 14), typically is within a building or is deployed for a cluster of network users that have similar needs, for example, an administration department, facility management department, or computer laboratory.

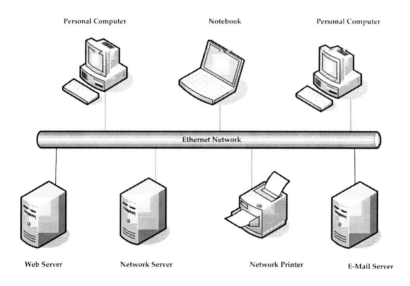

Figure 14 Local Area Network

A Campus Network, serving a university, corporation, or government, connects the LANs within each of the buildings on campus to provide a "seamless" network (Figure 15). Campus networks have connectivity between buildings provided via cable, although alternatives include the use of wireless and telecommunications circuits.

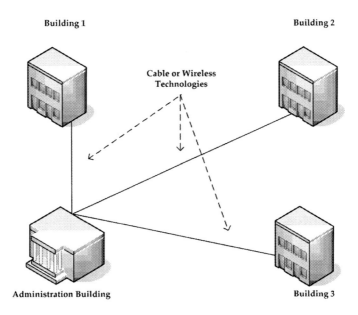

Figure 15 Campus Network

Data networks can scale to metropolitan areas and to wide areas, covering counties, states, continents, and even the globe. Connectivity between sites can be provided via point-to-point terrestrial (land lines) or non-terrestrial (such as satellite or wireless) telecommunications circuits, or the Internet.

A user's personal computer has its own processing, memory, storage, operating system, and software applications, and can operate independently of other systems or networks. However, connecting the PC to a network allows it access to other resources including printers, the Internet, and centralized databases or software. A LAN consists of four general device types:

- Users' Personal Computers or Other User Devices
  Typical desktop computers, laptops, and PDAs access a network through a cable, wireless device, or through a telecommunications circuit, like a dial-up telephone line. There are variations of the user devices and personal computers typically used on a network, such as "thin clients" (essentially a PC without any local storage), and network "appliances," devices geared to a specific application, such as Internet access and e-mail.

- Network Switches
  Network switches are the "traffic cops" for the network. When devices on the network (users and servers) transmit information over the network, the network switch determines the origin of the message or data packet, its destination, and the best path or route to get the message to its destination. The switches manage potential collisions on the network, and can provide a "virtual" network within the LAN for specific network traffic.

- Network Servers
  Servers are connected to LANs and provide a variety of resources to both network users and network administrators. A common application provides

network users connectivity to the Internet and shared Internet firewalls. Servers have many other uses including hosting web pages, hosting e-mail applications, print servers allowing users to share a printer, centralized databases or software applications, and network administration capabilities.

From a hardware perspective, servers can vary from a desktop PC to a mainframe computer. However, most servers are specialized with multiple processors, hardware redundancy, large storage capacity, specialized operating systems and software applications. They may be a desktop PC, or be installed in an equipment rack, or, for high density, may be a "blade" in a server chassis mounted in an equipment rack.

Integrating technology systems in buildings is analogous to a data network in that there is a "core" network where there is commonality in the way systems communicate and their physical connection. Outside of the core network are various devices and controllers from different manufacturers, with different capabilities and functions. This type of standardization has significant functional and cost implications, which are at the nucleus of the deployment and benefits of smart building technologies.

Voice Networks

In the past five years, wired telephone service within a building or an organization has undergone a technological revolution. For years, stretching back to the late 1970s, large telephone systems used a technology called "Time Division Multiplexing" (TDM). The market, during those years, was dominated by three large companies with systems open enough to connect to the outside world, but which had proprietary operating systems and core hardware.

Those systems, when deployed for larger installations, are referred to as Private Branch Exchanges (PBXs), and are essentially privately owned hardware that could connect and exchange calls with the outside world (Figure 16).

The market for telephone systems significantly changed in the late 1990s with the introductory of VoIP technology. VoIP essentially utilizes a data network based on the Internet Protocol to transmit voice. It encodes analog voice into digital data packets at one end and decodes the digital data packets into analog voice at the other end. As the marketplace changed, hybrid systems with both TDM and VoIP technology were made available. The hybrid systems are transitional systems allowing users to gradually evolve to VoIP technology.

41

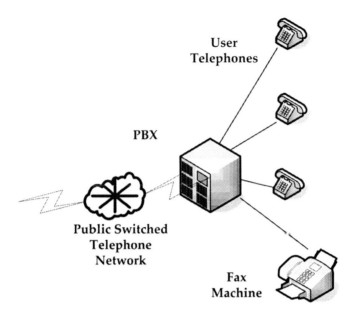

User
Telephones

PBX

Public Switched
Telephone
Network

Fax
Machine

Figure 16 TDM PBX Telephone System

VoIP has many advantages over the older TDM technology. A main selling point of VoIP is that as voice essentially becomes an application on a data network, organizations need only to deploy one network rather than separate networks for data and voice. This results in savings, primarily related to management, administration, and maintenance resources needed for the network, as well as some resource consolidation.

VoIP offers more functions and features through the true integration of voice and data networks. Traditional TDM telephone systems had attempted to integrate telephone systems and data networks, but the functionality of VoIP in this area far exceeds anything previously developed. For example, when a caller leaves a voice mail message for a VoIP user, the voice mail can be automatically digitized as an audio file, attached to an e-mail, and sent to the user. A user can listen to and/or file the voice mail message on their personal computer or other receiving device. Another example is a customer using a company's web page to obtain customer service; if the customer decides that they need to talk with a customer service agent, they can do so by just clicking on an icon and instantly engaging in a voice-to-voice customer service chat.

Many VoIP telephones are capable of being easily relocated, only requiring network or Internet access. Phones connected to an organization's VoIP system can move from an office to a home to a hotel room, always operating in the same manner. The VoIP system can track where the telephone is and forward calls, providing the appropriate calling privileges and features.

VoIP also has its issues and concerns, primarily with the ability to deliver voice transmission in real time. Traditional TDM systems essentially establish a circuit between the caller and the party being called, with the circuit reserved solely for that particular voice call. It is different in the VoIP world.

VoIP encodes voice into data packets. Data packets are then transmitted over the network, in which the data packets compete for network resources and bandwidth. The competition of the data packets for network resources can produce delays in the transmission of the data packets, even loss of or errors in the data packets. The result of such effects is the degradation of the voice transmissions. To compensate, the data network must provide a certain "Quality of Service"

(QoS) for the voice transmission. This is accomplished by the data network determining which packets are for voice transmission, prioritizing those packets, and dedicating network resources, through hardware or software, for the voice packets.

Another VoIP issue has been providing power to VoIP telephones. In a traditional TDM system, power is centralized and feeds each of the instruments attached to the system. In early deployments of VoIP, VoIP telephones had to be powered locally, at the location of the telephone instrument. This meant that not only did the telephone need a network connection, but that it also required a power outlet thus adding cost to VoIP deployments and making the moving, adding, and changing of instruments cumbersome. The market response was to develop a method to provide power to the telephone instruments from the network switches. It resulted in the IEE 802.3af standard, which set guidelines for providing power over a network cable. It is commonly referred to as "Power over Ethernet" (PoE).

Telephone users can use either analog or digital telephones. On traditional TDM telephone systems, standard analog telephones or proprietary digital instruments would be used. On VoIP systems, analog telephones can be use, with either an adapter at individual instruments or a network gateway which converts multiple analog telephones to the digital network. Other analog devices, such as fax machines, are serviced in the same manner as analog telephones.

Digital telephones for a VoIP system can connect directly to a network switch of a LAN, much like a desktop PC would be connected. Many digital telephones come equipped with a "mini data switch" built-in, allowing one network connection to be connected to the instrument, which serves both the telephone instrument and a desktop PC connected to the mini-switch. Unlike traditional TDM telephones, many VoIP telephone instruments come with the capability to browse the Internet or access an organization's network, including applications and databases.

Softphones are another option for VoIP systems. Softphones are software applications for a personal computer that essentially turn the PC into a telephone. Users utilize a headset or a USB-connected telephone to compliment the software application and have a fully featured telephone set. Software on the personal computers allows for on-screen dialing and access to a user's contact lists.

It is clear that VoIP will be the dominant future technology used in telephone systems. With telephone systems mimicking and riding on data networks, and utilizing similar standard cabling infrastructure and dominant network protocols, telephone systems can easily be integrated and converged into other smart building systems.

Audio Visual Systems

Audio visual systems are a complex topic as they can encompass scores of different types of equipment and material, multiple technical standards and rapidly changing technologies. Despite the complexity, and in many ways because of the complexity, smart building technology (standard structured cable system, Ethernet connections, and IP protocols) is taking hold in audio visual systems. This includes digitizing the traditional analog audio and video signals, and, more importantly, using the technology of data networks to control and manage audio visual systems.

There are a number of audio visual systems that can be deployed facility-wide. These include voice paging systems and intercoms. However, many audio visual systems are designed for the specific needs of certain rooms and spaces within a facility, such as meeting rooms and classrooms, rather than being facility-wide (Figure 17). The core parts of these audio visual systems are:

- Sources
- Processing and management
- Destinations (speakers and displays)
- System control

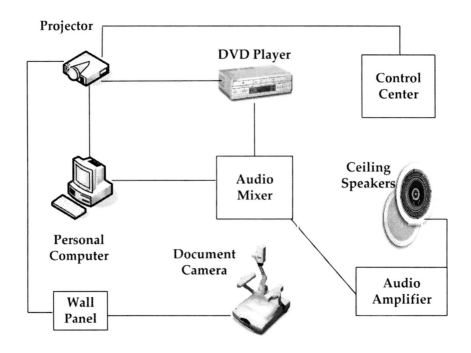

Figure 17 Typical Classroom Audio Visual System

Audio and Video Sources

The sources of audio are generally from microphones, electronic instruments, or programmed sources. The output from these audio sources is processed, adjusted, amplified, and fed to speakers.

Microphones convert acoustic energy (such as talking and singing) into electrical energy. Microphones come in a variety of types including handheld, wireless, lavalier or lapel clip-on microphones, headsets, ceiling and table mounted microphones. Electronic instruments, AM-FM tuners, audio CD players, cassette decks, MP3 players, and personal computers are also audio sources.

The visual part of an audio visual system also has several sources:

- Cameras
- VCRs, DVDs, VTRs, DVTRs
- Cable television
- Satellite television
- Broadcast/cable/satellite
- Document cameras
- Personal computers
- Data networks

The wide range of video sources also means a wide range of video signal types and methods with which they connect to an audio visual system. Some of the emerging video sources include electronic white boards and the Internet.

Personal computers are typically part of an audio visual system in meeting rooms, conference rooms, and classrooms, where presentations, often relying on Microsoft PowerPoint® presentation software, are used by the speaker or instructor. The personal computer serves as a video input to an audio visual LCD projector.

Document cameras have replaced the older, bulky overhead projectors. The document camera, which contains a video camera, is also typically connected to a LCD projector. Document cameras can handle a variety of materials, including X-rays and even 3-D objects.

Interactive or electronic white boards can be used to display an electronic presentation, but can also be used as an A/V source, where everything written or displayed on a white board can be recorded for electronic retrieval later. One application is in the classroom where an instructor utilizes the electronic white board and the content on the white board can subsequently be electronically retrieved by a student via a data network or the Internet.

Audio and Video Processing

Audio signals may be processed in a multitude of ways, with the most familiar being equalizers on many car and home stereo systems. An equalizer processes the audio signal by increasing or decreasing the low-, mid-, and high-level frequencies to provide the most pleasing audio sound possible. Other common audio processors include:

- Echo Cancellation
  This reduces the chance of sound output through a loud-speaker being re-amplified into the sound input. The re-amplification results in the well-known, squealing feed-back we have all heard at some time, such as when a microphone and a speaker are moved too close together.

- Limiters
  As the name implies, they limit the amplitude of the sound, restricting the volume to a predetermined setting.

50

- Compressors
  Compressors are like limiters, but instead of just limiting the loudest sound, they "compress" the audio signal to a certain volume range, avoiding both the loudest and the quietest sounds.

- Expanders
  These simply expand or increase the volume of the audio signal to a predetermined sound level.

- Gates
  Gates block or eliminate audio sounds below a certain frequency level.

- Automatic Gain Control
  This automatically controls the volume of the signal at a specific frequency level.

- Delay
  Delay of an audio signal is typically used in large venues where natural sounds can be heard after processed sounds, and delay is used to synchronize the two.

- Digital Signal Processing
  This is used with digital as opposed to analog audio and video signals.

- Distribution Amplifiers
  Distribution amplifiers take a single audio input, amplify it, and distribute it to multiple audio outputs.

In many deployments of audio visual systems, more than one sound source exists, and an audio mixer is needed to combine or mix the multiple signals. For example, if there were three panelists speaking at a conference, each with their own microphone, the mixer would combine and adjust the individual signals as needed. After the processing and combining, the audio signal may be routed to a set of speakers, possibly using an amplifier in between. Mixers are used in recording and broadcast studios (mixing boards), but also can be part of a public address system and audio visual systems for large meeting rooms. Mixers have various features such as the ability to add effects to the sound, and connection for a personal computer to provide enhanced equipment management.

Like audio signals, video signals also need to be processed, adjusted and "groomed". The processing of video signals is however significantly different than audio as the amount of information for video signals is magnitudes beyond that of audio. Video processing may involve amplifying, or adjusting the timing, color, brightness, or contrast of the signal. Video processing equipment can include time base correctors to maintain the integrity of the signal and video processing amplifiers.

Audio and video signals both need devices to switch and route signals. An audio visual switcher is a "traffic cop" type of device with the ability to connect a signal input to a single output connection. A matrix switcher or router is just multiple switches, with the capability to switch multiple inputs to multiple outputs. Video switches are primarily used in video production and video distribution. For example, a live event such as a football game can be covered by several cameras, and a video production switcher allows production personnel to easily switch between cameras. A matrix switcher may be used with a video distribution system within a building, allowing for different video sources to be routed to different areas or rooms within a facility.

The amplification of audio and video signals increases the voltage and power for further distribution of the signal before delivery to its destination. Video distribution amplifiers typically take a signal output, amplify it to maintain signal quality, and distribute it to multiple outputs. A room having multiple displays, where one video signal is to be distributed to all available displays will utilize a video distribution amplifier.

Speakers and Displays

Loudspeakers perform the opposite function of a microphone, converting electrical energy back into acoustical energy. They vary in the range of audio frequencies reproduced and the dispersion pattern of the sound. There are two types of loudspeakers, cones and compression drivers. Cones can be used for reproduction of the bass, mid and high frequency spectrum, while compression drivers are used only for mid and high range frequencies. Since no one loudspeaker driver can accurately reproduce the entire range of frequencies, several drivers may be utilized in a single speaker enclosure.

53

A video display device is one that presents information visually. Video displays come in a wide variety of sizes and core technologies, and are utilized in a wide variety of environments. Selection of a display is based on the dimensions of the viewing area, lighting, and the type of content and materials to be displayed.

Display devices can be front or rear projected. Front projection is the preferred display in presentation environments (such as classrooms and meeting rooms) because it is less expensive than rear protection and requires much less space. The viewing image for the audience is projected onto a front screen, which reflects the image.

Rear screen projection is preferred in bright, well-lighted environments. Basically, light from a rear screen projector is transmitted through the screen. The screen is made of material that transmits the light with relatively little distortion.

Rear and front screen projections can use Liquid Crystal Display (LCD), Cathode Ray Tube (CRT), or Digital Light Projector (DLP) technology.

Major types of video displays include:

- Plasma Screens
  Plasma screens are basically a network of chemical compounds, called phosphors, contained between two thin layers of glass that, when excited by an electric pulse, produce colors, light, and a picture. The picture of a plasma screen is bright and rich with color. Screens can be as large as 80-inches. Plasma screens are thin, lightweight, have a wide

viewing angle, and offer good pictures under normal room lighting conditions.

- Digital Light Processing (DLP) Television & Projection
  DLP™ technology is based on a semiconductor invented in 1987 by Texas Instruments. The basic technology in a DLP monitor is millions of microscopic mirrors that direct light toward or away from pixels. The pixels control the amount of light reflected off of a mirrored surface. DLP monitors have excellent color reproduction and contrast, and are lightweight. The monitors are deeper or thicker than plasma and LCD monitors, because DLP uses a lamp. DLP technology is used in many rear screen projectors.

- Liquid Crystal Displays
  LCDs use a florescent backlight to send light through its liquid crystal molecules. LCD monitors apply voltage to the pixels, to adjust the darkness of the pixels, thus preventing the backlight from showing through. Many LCD displays double as computer displays. An LCD display can be as large as 55-inches. LCDs are very thin, lightweight, and have good color reproduction and sharpness.

- Cathode Ray Tube
  CRTs use a vacuum tube which produces images when a phosphorescent surface is excited. The dot pitch (DPI or dots per inch) of a monitor (the distance between the dots) is a measure of the quality of resolution. CRTs have a large dot pitch, which mean a lower resolution quality, but CRTs

produce a brighter picture. CRTs have excellent color and contrast, but tend to be big and heavy.

Audio Visual Control Systems

A control system has to simultaneously manage all the various components of the audio visual system. A presenter or user may want to lower a ceiling mounted projector, start the DVD, or perform other related tasks. Some elements of an audio visual presentation system that may need to be controlled include:

• Projectors	• Monitors
• Camera	• VCR
• DVD	• Lighting
• Screens	• Room Access
• Curtains, Drapes	• Volume
• Alarms	• HVAC

Local control of most audio visual systems may be a series of control buttons and switches located on a wall within the room, a wireless device controlling a projector, display, or other component, or a touch screen on a computer. Traditional system controls may include relays, remote controls, and proprietary manufacturer controls. Increasingly, however, the control system for audio visual systems is a personal computer, PC software, and the use of a data network utilizing Ethernet and IP protocols.

Although digital audio and video content can be transmitted over a data network, most of the evolution of audio visual systems to data networks has been in the administration and control of the components. While some components of an AV system may have direct network connections, such as projectors, many other components do not. However, projectors, cameras, VHS/DVD players, motorized projection screens, lighting, and windows shades or curtains are AV devices that can be typically be controlled in an IP network.

Some of these components may connect to an "Ethernet Interface" or bridge device (Figure 18), which interfaces the various inputs and outputs from multiple components onto a standard Ethernet IP network. These systems add a network "on top" of an audio visual system to compensate for some of the components shortcomings in connecting to the network directly.

The capability to administer an audio visual system through a standard smart building technology allows functionality previously not available. This includes remote control of the system over the network, enterprise-wide asset management, preventative maintenance, content delivery, component software upgrades, and more. Remote monitoring allows a service technician to use a web browser to access and monitor all the components of an audio visual system. The technician could turn a projector, or other component, on or off, run diagnostics on the equipment, centrally control all displays, and the like. Remote communication can be two-way. For example, technical personnel can be notified via email of an alarm or a maintenance event for a particular AV component, such as a burnt out projector bulb.

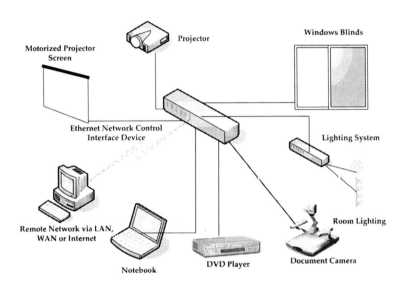

Figure 18 AV System with Ethernet Interface

Beyond system administration, management, and monitoring through data networks, audio visual systems have increasingly incorporated the digital creation, sourcing, transport, display and storing of audio and visual signals. The wide variety of components and the types of cable interconnections means audio video system's adoption of smart building technologies is more complex than other technology systems.

Video Distribution Systems

Distribution of video programming is related to but different from audio visual systems. The best example of video distribution is cable television, where a service provider is distributing video programming to its customers.

Video distribution is different for different types of buildings. For residential buildings, such as houses, apartments, condominiums, and dormitories, video distribution is fairly dense because most television viewing takes place in one's home. In most commercial and government buildings, distribution of video is primarily in common areas, building entrances, cafeterias, meeting rooms, elevators, assembly rooms, classrooms, and so forth.

Most video within buildings is currently being distributed using a technology developed decades ago, commonly referred to as CATV (Community Antenna Television) or RF (Radio Frequency). The technology takes multiple analog video signals and "modulates" or places them on different radio frequencies, carried by a coaxial cable (Figure 19). Many large cable television service providers and others have moved away from the use of coaxial cable to some hybrid of fiber optic cable and coaxial cable or, in some areas, just fiber optic cable, to transmit the video RF signal.

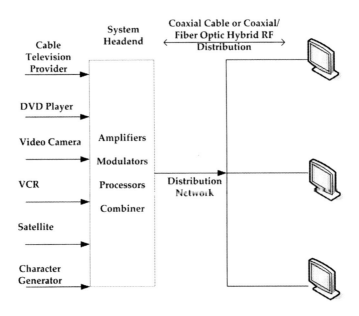

Figure 19 Traditional Video Distribution

Video distribution, like other technology systems within a building, has evolved and is moving to a standard cable infrastructure (unshielded twisted pair copper and fiber optic cabling). Systems can also take a typical analog video signal, convert it to a digital signal, and transmit it over data networks conforming to the IP protocol. Therefore, in many ways, video distribution is following the evolutionary path that voice networks have taken with VoIP.

Video distribution within many buildings is simply a retransmission of programming provided by the local cable company. More sophisticated systems may selectively filter or block some of the programming from a local cable company, or insert local sources of video from a DVD, camera, or character generator.

An emerging use of video distribution in buildings is digital signage, where the video distribution system is conveying information or advertisements through text, video and sound. The programming content can be easily and remotely changed and updated to provide real-time information or pre-determined delivery of previously stored data.

Beyond the basic video distribution lies many forms of video distribution. Video distribution can be point-to-point or point-to-multi-point, it can be one-way (unidirectional, much like the delivery method of cable television into a house) or two-way (bidirectional, where video signals are sent upstream to be retransmitted downstream). Some of the applications include:

- Media Retrieval
  This is an application where video programming is stored either in digital or analog format, and a system administrator or authorized individual users can schedule, or "call-up," in real-time, specific video programming. For example, a mathematics teacher, possibly with the assistance of the librarian or media director, can call-up a program entitled "Practical Applications of Algebra" for viewing by the teacher's 2:00 PM algebra class. Video on demand, primarily for movies, is also becoming more mainstream with some advanced cable television systems.

- Video Conferencing
  Video conferencing is an application primarily for corporate and organizational meetings, although it is a growing application for personal computer users with a web camera. The application addresses the cost and potential ineffi-

ciency of live meetings. Video conferencing can be augmented by graphics transmissions between locations. Video conferencing is bi-directional and the audio quality between locations is critical to its success.

- Distance Learning
  Distance learning is somewhat a specialized form of video conferencing. Typically, distance learning has an instructor in one location and a class at another location. The application may also be augmented with graphics transmissions. Classrooms may have a series of speakers and microphones. The application can include subsystems, such as remote response systems, where students have a keypad with audio capabilities thus facilitating communications between students and the instructor, and even testing.

- Live feeds from video cameras
  We see this everyday when watching TV news programs. In a building or campus environment, it may be a live camera feed from a CEO's or Principal's office to employees or students gathered in meeting rooms, classrooms, and common areas.

Video Display and Viewing

In 1941, the U.S. National Television Standards Committee (NTSC) proposed the original analog television format for black and white television, following in 1950 with a backward compatible standard for color television. Many of us still watch television in the 1950 standard.

Analog video is a group of still images or frames. Thirty of these frames are broadcast each second giving viewers a visual sense of motion. The quality of the resolution of any video transmission is determined by the number of horizontal lines displayed on the screen and the number of tiny dots, or pixels, that emit either red, blue, or green light to create the picture we see on the screen. The total number of pixels is the measure of the quality or resolution of the picture.

The 1941 analog standard called for 480 horizontal lines of resolution (additional lines are reserved for synchronization and information such as captioning), 720 pixels per horizontal line, and a technique called interlacing. Interlacing means that only half of the horizontal lines are used in creating each frame; that is, even number lines are broadcast in one frame and odd number horizontal lines are broadcast in the next frame. The total resolution of a screen can be calculated by multiplying the number of pixels in each line by the number of lines on the screen (720 pixels per horizontal line x 480 lines = 345,600 total pixels).

There are currently 18 different digital television formats set by the Advanced Television System Committee (ATSC) standard. These are formats for televisions and displays that can be grouped as follows:

- Standard Definition (SDTV)
  This is basically the analog television standard delivered as a digital signal, with three variations of the formats dealing with the number of pixels, the shape of the pixels, and using progressive scan (which uses every horizontal line in creating a frame) rather than interlacing.

- Enhanced Definition (EDTV)
  EDTV is like SDTV, with the main difference being it uses progressive scanning rather than interlacing. There are nine variations of EDTV dealing with the number and shape of pixels, as well as the aspect ratio.

- High Definition (HDTV)
  HDTV offers the highest quality resolution and uses a 16:9 widescreen aspect ratio only. HDTV has six variations addressing lines of resolution, number of pixels, frame rates, and interlacing.

Digital Video Transmitted Via a Data Network

If one were to digitize the traditional analog video signal, assuming each pixel needed three bytes of information to indicate color, the digital signal would need a transmission rate of over 8 Mbps, a high bandwidth network requirement that is still beyond the vast majority of data networks. To reduce the size of digital video, several techniques were deployed: the number of frames per second was decreased, the sizes of screens were reduced, transmission for pixels that did not change frame-to-frame was eliminated, compression techniques were used, and so forth. The result was the development of hardware or software codecs (<u>co</u>ding and <u>dec</u>oding devices) that compressed analog video and decompressed digital video signals. Codecs were developed around a series of compression standards, such as the H320 and H323 standards.

Video can be digitized in several formats, but typically uses an MPEG (Motion Pictures Expert Group) standard for the encoding and compression of digital multimedia content. MPEG-2, or its latest incarnation, H.264, provides compression support for the TV quality transmission of digital video. Other standards address channel change signaling and video on demand.

Delivery of video over a standard data network can be performed in several modes:

- Multicast mode - A method in which programming can be sent to multiple devices at the same time
- Unicast Mode - A method in which programming can be sent to just one device
- Broadcast Mode - A method in which programming can be sent to all open devices at the same time.

A typical deployment of video distribution within a smart building involves (Figure 20):

- Digital television transmitted over a TCP/IP-based data network
- Digital video distribution which transmits and receives MPEG video streams
- Distribution and transmission of digital video in a broadcast, multicast, or unicast mode
- A media retrieval system, allowing authorized users to schedule and stream video programming on demand
- An extensive management and reporting system

- Inputs to the digital video distribution system including live or recorded video, encoded analog video, cameras, and character generators
- Digital signage

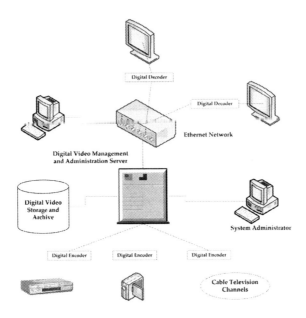

Figure 20 Digital Video Distribution

Video distribution will continue to penetrate, evolve, and be delivered through standard smart building systems, and it has reached the market acceptance and advanced functionality that VoIP has achieved.

Fire Alarm Systems

Fire alarm systems are a primary life safety system for every building. Properly deployed, a fire alarm system reduces the probability of injury or loss of life and limits damage due to fire, smoke, heat and other factors.

Because of their criticality, the codes, regulations, and standards affecting the design and installation of fire alarm systems are wide ranging and detailed. Their design and installation must involve qualified, licensed experienced professionals and, more importantly, the coordination and approval of the local authority having jurisdiction (AHJ).

Even given their life safety nature, fire alarm systems can, and should, be integrated and converged with other systems within a smart building. The fire alarm system initiates sequences of operations for other building automation systems to facilitate evacuation from the building and containment of the fire, such as:

- Signaling the HVAC system to restrict and contain smoke, heat, and fire through its dampers and fans

- Using the access control system to clear a path for building evacuation by opening doors, unlocking secured doors and releasing powered exterior doors
- Using the access control system to contain and prevent the spread of fire and smoke by closing interior doors
- Triggering emergency power for the fire alarm system and related systems operation, exit signs, and lighting for building exit routes
- "Capturing" the elevator and shutting down its operation.

The fire alarm system must communicate with and control the system components, and it also must communicate with off-site facilities and organizations, such as the fire department and emergency services. The networking of the fire alarm system, like other systems, is accomplished with a cable infrastructure and communication protocols.

The reliability of a fire alarm system is partially dependent on the system cabling. Both the National Electric Code and National Fire Protection Administration have specific guidelines to ensure the system's proper operation. Traditional fire alarm systems may use unshielded twisted pair and fiber optic cables between panels. The cable used for a fire alarm system can be the same structured cable infrastructure used by other smart building systems, thus allowing fire alarm system integration with those systems.

The communications protocols used by major fire alarm systems manufacturers typically conform to a BAS protocol, such as BACnet or LonTalk, while some system have introduced some use of the IP protocol between major components.

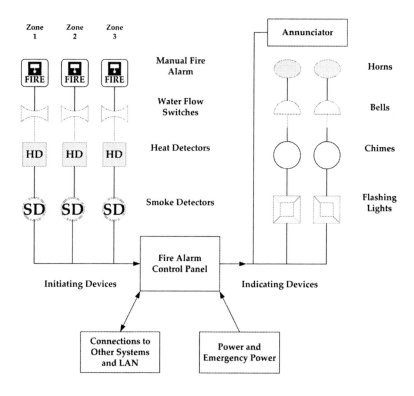

Figure 21 Basic Fire Alarm System

Fire Alarm Control Panel

The heart of a fire alarm system is the primary fire alarm control panel (FACP). The FACP monitors system integrity and starts all sequences of operation for the detection, suppression, and notification of the system. FACPs are typically microprocessor based with software for communications, processing, and decision making. FACPs gather data from system detection or indicating devices, process the data, and then act on the data if need be, managing system alarms, suppression systems, as well as other building automation, security, and life safety systems. FACPs can also detect failures within the system that require repair and maintenance.

Depending on the size of a building or the number of buildings on a campus, an overall system may have one central control unit or may have distributed control units networked together to provide a unified and transparent system. Field or remote control panels can be networked to a central panel. The relationship between the central panel and the field panel can be "peer-to-peer" or "master-slave." Field panels can be deployed for narrower, more specific functions, such as simply supplying power to alarm devices, or used as a remote annunciator.

System devices assigned to a zone can be connected to the FACP either through a direct connection, a bus, or a loop topology. The National Fire Protection Adminstration designates the performance of these connections or circuits. Class A circuits must be able to transmit a signal even with an open or ground fault on the circuit, while Class B circuits need not. The loop connection can be "fault-tolerant," thus allowing continual operation if there is a break in the cable or a device failure.

These connections to the FACP and the devices on the connections can be either "addressable" or "supervised." Addressable means the FACP can communicate to a single device or a group of devices depending on the functions required. Supervised devices are monitored to ensure that they are still connected to the FACP and are operational.

Regardless if devices are supervised or addressable, the system is required to continuously monitor their status. For addressable devices, monitoring is done by the polling of individual devices. For non-addressable devices, monitoring is accomplished by current sensing, where the FACP provides a small current which, if interrupted, indicates trouble, such as a failed or removed device.

Annunciator Panel

A major component to the system is the annunciator panel attached to the FACP. The annunciator provides visual and audible indications that an alarm has been initiated, and provides the location of the alarm. It may also identify the functions that could affect the fire and the building occupants in the area. A basic annunciator panel may have an alphanumeric display, and switches to acknowledge and silence the alarm. More sophisticated annunciator panels have a monitor or include a personal computer with graphical displays, displaying floor plans of the building and the location of the alarm.

Fire Detection

Fire consists of smoke, heat, and light. The system components that detect the fire and initiate an alarm monitor one or more of the fire's characteristics. The detection components of a fire alarm system are typically located in ceilings, HVAC ducts, mechanical and electrical areas, equipment rooms, and so forth. They include, but are not limited to:

- Pull Stations, where someone sees a fire and pulls the fire alarm
- Thermal Detectors, which sense a rise in temperature or the high temperature of a fire
- Smoke Detectors, which sense vapors of small particles of carbon matter generated by burning
- Flame Detectors, which sense radiation and visible light from a fire
- Fire-gas Detectors, which sense gases, such as carbon dioxide and carbon monoxide
- Air Sampling Fire Detectors, which are the most sensitive type of detection available. They are used in high value and critical environments such as churches, clean rooms, hospitals, museums, and communications and network equipment rooms. This system typically uses tubes to continually draw air samples to a highly sensitive detector. The system can detect a pre-combustion stage of a fire, prior to any visible smoke or flame.

Suppression Systems

Fire suppression systems include:

- Wet sprinkler systems, which may have various switches and flow detection equipment that is monitored and managed
- Dry sprinkler systems, which may have pressure switches that are monitored and managed.

Fire suppression systems can also have monitoring and supervision equipment associated with the fire suppression system, which are in addition to similar components in the overall fire alarm system.

Notification Devices

Once a fire is detected, a building's occupants must be notified so they may evacuate the building. Fire notification devices use audible, visual, or a combination of audio and visual signaling. These devices are typically DC powered so they can operate on backup batteries, and they must also adhere to product compliance as determined by Underwriters Laboratories' testing and listing guidelines for use as a fire alarm notifier. Fire alarm notification devices include, but are not limited to, the following:

- Bells
- Chimes
- Horns
- Speakers
- Strobes, including strobe lights combined with other devices

Monitoring

Fire alarm systems have two classes: (1) a protected premise system, which is a single building or campus of buildings under control of one owner, protected by a single system, where the system is monitored locally or remotely by the owner, and (2) a supervised station system, which is much like a protected premise system, except the system is continuously monitored by a third party security or central monitoring company. Monitoring is addressed in the following ways:

- Local Monitoring – When activated, a local alarm announces an alarm to the area it covers.
- Remote Monitoring – When activated, a local alarm will be monitored remotely by building or campus personnel through a communication network.
- Supervised Station – When activated, a local alarm will be monitored by an off-site company that provides recording, supervision and management of the fire alarm system.

The fire alarm system is the smart building system requiring the most interaction with other smart building systems. The interfacing, integration and convergence of the fire alarm system with other systems are not only desirable but a smart building requirement.

A fire alarm system can integrate with the structured cable infrastructure used for all other building systems. Fire alarm systems are starting to deploy some use of IP communications protocols, primarily between fire panels. There is still substantial reliance on communications protocols developed for building automation systems (BACnet, LonTalk, etc.). However, these protocols can be easily routed to an IP system, and embellishments of some protocols, such as BACnet/IP, are evolving as the dominance of the IP protocol is becoming more widely recognized.

Video Surveillance Systems

Video surveillance systems, also known as closed-circuit television systems (CCTV), are one part of a larger facility's security plan and deployment, which should also address physical and operational aspects of security.

Deployment of video surveillance systems must take into consideration some legal aspects, mainly a person's right to privacy and the presumption of security. For example, one should not put a camera in an area where a person should expect privacy, nor should one deploy "dummy" cameras so as to provide the perception of security monitoring, when, in fact, none exists.

Much like video distribution systems and the broader electronics marketplace, the technology for video surveillance has, for decades, been based on analog technology. However, this is rapidly changing.

Video surveillance systems basically perform five functions:

- A camera captures a picture or video image
- The video is transmitted back to a security control center
- The video is processed

- The video is recorded
- The video is viewed on a monitor.

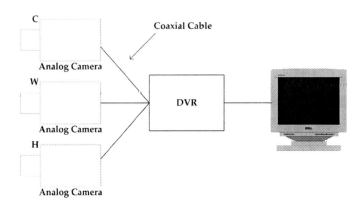

Figure 22 Typical Video Surveillance System

Video Capture

The picture or video is captured via the camera and lens assembly. Surveillance cameras are either fixed cameras, which provide one view, or pan/tilt/zoom (PTZ) cameras, which provide several views and can be controlled remotely from a security control center.

The panning and tilting of a PTZ camera is accomplished by the camera mount, not the camera itself. The one drawback of a PTZ camera is that while it views one specific area within its range, some event could take place in another part of its panning area and go unnoticed. To compensate, PTZ cameras can pan in a pre-defined pattern or can be interfaced with other systems, such as access control, so that it automatically pans to a specific area if triggered.

Cameras come in a wide variety of resolutions, and some have advanced features to control and adjust lighting. Placement of security cameras must consider the camera, the lens, the lighting, power and, most importantly, the security need.

Variable lenses are the most commonly used camera lenses allowing for adjustment in field. Zoom lenses can be adjusted remotely from the security control center. Color cameras are typically used in most video surveillance systems, although black and white cameras actually are better in low light situations. Combination cameras, call "night/day" are available for exterior uses where lighting varies regularly. Camera mounts and housings for cameras are available for a variety of placements and aesthetic tastes. Housings are also available for special environments, such as prisons or laboratories.

The true revolution in cameras has been the digital surveillance camera coupled with an IP network connection. A networked camera digitizes the video (typically into a MPEG or motion JPEG format), compresses the digitized video, creates a data packet of the video, and transmits the video over a standard data network. The cameras may have the capability to store video, to buffer the video if network traffic is heavy, or to process video. In additional to the digital video output, the network connections can transmit PTZ signaling, audio, and other control and management commands. The cameras can also be powered from a central location utilizing PoE.

Video Transmission

Transmission of the video signal captured from a surveillance camera to the security control center typically has used coaxial cable, the traditional cable for analog video. With changes in the technology, more installations are using unshielded twisted pair copper cable and fiber optic cable. Unshielded twisted pair is even being used with analog cameras, with baluns (an interface between balanced signals and unbalanced signals) or a manufacturer's proprietary technology, which may allow signaling over long distances.

With IP cameras, transmission is accomplished through un-shielded twisted pair cabling, as part of a structured telecom-munications cabling system. Fiber optic cable is utilized for exceptional long cable runs or for exterior cameras, where lighting protection is a concern. The distance between the camera and the head end equipment, as well as cost, security of signal, and resolution, may be considered in selecting the physical transmission media.

Wireless transmission can be used for cameras where cable is impractical or costly. Wireless can be deployed rapidly, but it may require power and a line of sight between locations and be susceptible to interference. Wireless technologies include "Wi-Fi," infrared, microwave, and Free Span Optic (FSO) systems.

Video Processing

Video processing codes or encrypts video signals from the surveillance cameras. The processing allows multiple cameras to be displayed on a single monitor or multiple camera views to be cycled on a monitor. Two products, both of which are remnants of analog video systems, have been typically utilized to accomplish this: a cross-point matrix and multiplexer.

The basic functionality of the older cross-point matrix is to switch camera feeds to different outputs. It takes the feeds or inputs from the video surveillance cameras, and "binds" any of the input to any output or multiple outputs.

The functionality of the older video multiplexer is to take multiple video feeds and combine them into a single video feed. This allows multiple cameras to be viewed on a single monitor. Multiplexing does this by lowering the resolution of each video feed and rescaling it into a screen, typically showing the views of 4, 9, or 16 cameras.

In a digital system, with IP cameras and a network server replicating a cross-matrix and multiplexer, the system functionality is significantly enhanced through its use of a software-dependent approach, rather than the hardware-dependent approach of traditional analog systems.

Recording

One of the first components of video surveillance systems to go digital was the digital video recorder (DVR), introduced to replace the older, tape-based video cassette recorder (VCR). The VCR worked with analog video and typically sat behind a multiplexer, which made multiplexing several cameras onto one video cassette tape possible.

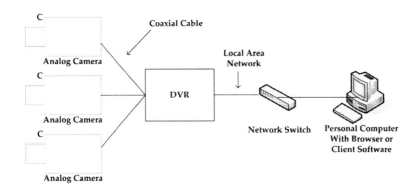

Figure 23 Video Surveillance - Networked DVR

DVRs were first introduced into traditional analog systems. DVRs can digitize video inputs from analog video surveillance cameras and also have the functionality of a multiplexer. Hard disks are the recording media for DVRs, thus eliminating the need to consistently change out or store tapes.

DVRs evolved further to include an Ethernet network connection, thus allowing the digital video to be transmitted over a data network, and opening up the possibility of viewing video remotely and viewing video through a web browser. Digital recording brought new functionality to video surveillance, such as the capability to detect motion in a picture, to record at different frame rates based on the detection of motion, to view video while recording video, and more.

The next digital evolution of video recording occurred when the DVR was replaced with a video server; that is, a data network server with video management software. In this arrangement, cameras, analog or digital, connect to the server and the server connects to the network. The video server uses standard data network equipment and becomes the center-piece of a video surveillance system. Because the server is on a network, it opens up a wide variety of functions for recording, storing, viewing, and administering the system, either on the network or off-site.

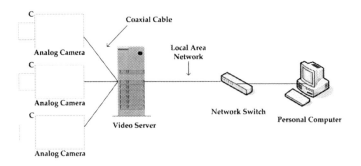

Figure 24 Video Surveillance System with Video Server

A major advantage of the video server is that the functionality of the system is derived from software rather than hardware, as was the case with older analog video surveillance systems. For example, an administrator or an authorized user can determine and set the number of inputs, the resolution required, and so forth for viewing multiple cameras or video on a monitor or screen.

Monitoring

Monitoring is categorized as the viewing and analyzing of video from the surveillance camera(s) and, if warranted, taking other security actions based on the analysis. Traditionally, this has been accomplished in a security control room with banks of monitors and security personnel. The viewing can be live video feed or recorded video. CRT technologies, which have traditionally been used in these control room environments, are being replaced with LCD and plasma monitors.

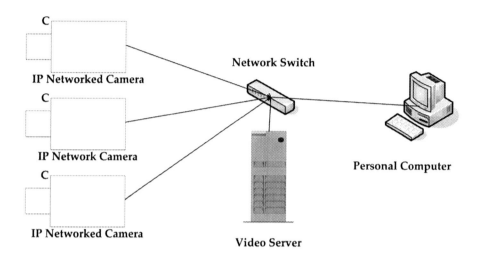

Figure 25 Full Digital Video Surveillance System

More importantly, the digital transformation in video surveillance has provided a variety of ways for monitoring to take place. Authorized users with an Internet connection and a desktop, laptop, or PDA can access the video surveillance system remotely and view video. Or, the system can be setup so that when an event is detected on a camera, an image is transmitted to a local or remote device of specific security personnel.

In addition, as the technology systems in a building are integrated and converged, separate control rooms, such as a security or data network control rooms, can be integrated into a more comprehensive network operations center, where all smart building systems are monitoring and administered.

Digital video surveillance systems using standard structured cable and IP and Ethernet protocols allow the system to easily become part of a smart building. They also demonstrate that digital systems provide increases in system flexibility, functionality, and scalability.

Access Control Systems

Access control systems have become more important as security has become more important for buildings. The basic and typical building access control system operates so that a person presents a card to a card reader for a particular door and, based on the information on the card and the system parameters for the person, door, and facility, the system either unlocks the door, allowing the person to pass through, or refuses entry.

The access control system is also important for life safety and is generally interfaced to the fire alarm system to facilitate building egress during evacuation. Access control systems must interface or integrate with several other smart building systems (video surveillance, HVAC, and others) as well as share data with business systems, such as human resources and time and attendance.

The marketplace for access control systems is very fragmented, with an array of manufacturers and suppliers, none of which is currently dominant. The result is a marketplace slowly moving away from proprietary systems and embracing industry standards. However, access control can and does utilize standard cable infrastructure and standard open protocols that are the foundation of smart buildings.

The basic components of an access control system include:

- A central host computer or server
- Control panels or system controllers connected to the host computer
- Peripheral devices, such as card readers, door contact, sirens and sensors connected to the control panels

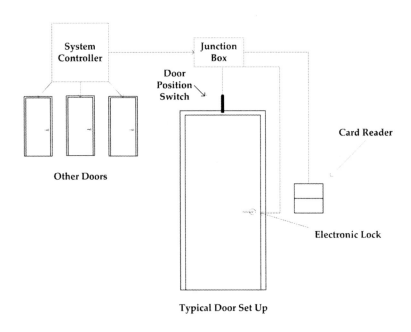

Figure 26 Access Control at Typical Door

Server or Host Computer

The server or host computer houses the operating parameters and the database for the access control system. The host computer is networked and communicates with the control panels, which collect data from the peripheral devices they serve, regarding events and alarms.

The connection and communications between the control panels and the host computer generally occur using one of two methods. One is through an RS-232 or RS-485 communication from the host computer to the control panels. Typically, this is accomplished by looping or "daisy chaining" a connection from control panel to control panel, with all control panels accessing the host computer through the first control panel on the loop.

The second and emerging method is to have the host computer and the control panels on a network using IP protocols and standard structured cabling.

The user database for the access control system has profiles for every authorized user relating to levels and parameters. This database can be centralized in the host computer or distributed among the control panels, with the host computer still having the complete database. This distributed network architecture is used to "push" the system decision making out to the local control panel, reducing the communications traffic to the host computer.

For instance, when a person arrives at a door and presents an access card with encoded credential information to a card reader, that information is then passed to the local control panel. In a centralized database system, the control panel passes the information on to the host computer, with additional information regarding the location of the door and the time the card was presented. The host computer then verifies the information and compares the access level to the door location and the time of day. Upon verification, the host computer sends a command to open the door to the control panel and the person is allowed access to the building.

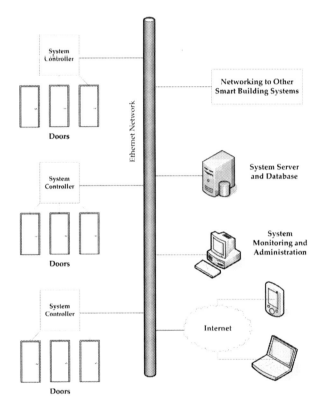

Figure 27 Networked Access Control System

In distributed database architecture, a control panel with onboard memory has the database for the doors and locations it monitors.

The control panel verifies the credential information without communicating to the host computer. The control panel later sends all transaction data to the host computer for system archiving. This distributed method can operate even if communication with the host is lost, including storing and buffering transaction data for sending to the host computer when communication is restored.

In addition to the database controlling cards and access levels, the host computer is capable of producing administrative reports and creating users' badges. The badging system at the host computer typically includes a camera, backdrop, and badging printer. A secondary badging station may be established in an organization's Human Resources department or elsewhere, where employees, visitors, students, and others are entering or exiting the organization.

The software on the host system has features to facilitate the security operations. It may have the ability to supply an automatic notification, allowing security personnel to be notified via a page or e-mail when a particular person is attempting or has been granted access to a facility. Or, if the entrance to a building has a guard, the guard may use a personal computer displaying information and an identification picture automatically when a person enters the building using an access card, allowing the guard to compare and confirm that the person using the card is indeed the authorized card holder.

Control Panels

Control panels are usually enclosed printed circuit boards with connections to all peripheral devices in their area. These peripheral devices may include door hardware (such as a card reader, door position switch, or door strike) and other inputs and relays as required. The control panel manages the peripherals devices and communicates between the host computer and the peripheral devices. The control panels:

- Consolidate all connections to peripheral devices
- Provide power, as needed, to peripheral devices
- Manage peripherals when communications to the host computer is absent or when acting in a distributed manner

Although control panels use a standard network connection and are able to communicate with the host computer or other control panels over the network, the signal methods of a control panel may be proprietary.

Most peripheral devices connected to a control panel do not have a lot of information to communicate. The device is either "on or off," or "open or closed." These devices (door contacts, requests to exit, door lock, auxiliary outputs and inputs) communicate with a simple dry contact. For example, when the system detects a "short" across the contact, which indicates an event (for example, the door is open), the system can take the required action, if warranted. The connection between the control panel and the card reader is more complex then just "on or off", entailing more information flow. The card reader typically uses the "Wiegand" industry standard to communicate this information.

The control panel communicates its status and recent access transactions to the host computer. The host computer or access server initially configures the control panel, may continuously synchronize the card holder data base, and performs monitoring, management, and administrative functions.

Peripheral Devices

The basic door-related peripheral devices include:

- Door Contacts
  Door contacts or door position switches are nothing more than an electromagnetic connection monitoring whether the door is open or closed.

- Request-to-Exit (RTE)
  This device is inside a controlled door and detects a person approaching the door who wants to exit, and allows the person to exit without setting off an alarm.

- Electrified Door Hardware
  These are components within the door or doorframe allowing the door to be automatically locked or unlocked.

- Card Readers
  There are a variety of access card readers, with the differences being related to the technology and the interface to the person requesting access to a facility. There are magstripe, swipe card, and insertion card readers, all of which use the same technology. Proximity and MIFARE readers are both "contactless" technology, operating at different frequencies, with MIFARE considered a "smart card" technology. There are also biometric readers, which uses biological information to verify a person, and involve such identification measures as fingerprint scans, face scans, retina scans, iris scans, hand structure, voice identification, and other methods.

Other input devices that may be monitored or supervised by the control panel include:

- Push Button - Primarily used for Americans with Disabilities Act (ADA) applications
- Panic Button – Used to notify security of an emergency
- Glass Breaks Sensors – Notifies security of a breach to windows
- Motion Detectors – Typically used in hallways
- Key pad – Key pads receive a code from a person, and upon verification, allow access

The host computer or control panel also has output relays, allowing the access control system to interface with other smart building systems (like lighting control or HVAC) or a local or remote annunciator.

While the cabling for the door hardware (door strike, card reader, door contact, and request for exit), is typically not part of a structured cable system (typically 2 to 4 shielded twisted pairs, with cable gauges varying from 16 to 22), access control systems have embraced structured cabling and Ethernet for system controllers and the centralized server. In addition, the requirement for the system to share databases with business systems has pushed the marketplace into offering databases that are compliant to SQL and ODBC standards.

Heating, Ventilating and Air Conditioning Systems

Heating, ventilation, and air conditioning systems (HVAC) maintain the climate within a building. They control the temperature, humidity, air flow, and the overall air quality of a building. A typical HVAC system brings in outside air, mixes it with air returned from or exiting the system, filters the air, passes it through a heating or cooling coil to a required temperature, and distributes the air to the various sections of a building.

The HVAC system not only makes the building comfortable and healthy for its occupants, it manages a substantial portion of the energy used and the related cost of the energy for the building. In maintaining the building's air quality, the HVAC system must respond to a variety of conditions inside and outside the building (including weather, time of day, different types of spaces within a building and building occupancy) and do so while optimizing its operation and the related energy usage. The HVAC system is also critical in controlling smoke in the event of a fire.

HVAC systems in commercial and institutional buildings are very different than those used in typical residential housing. The larger buildings have a higher density of people, lighting, and other equipment, all of which generates more heat. The result is that air conditioning, or the recirculation of air, becomes more important than providing heat, dependent on the local climate. Although there may be a centralized HVAC system in commercial and institutional buildings, different sections of large buildings have different HVAC needs or thermal loads depending on how the space is utilized.

A HVAC system having one control thermostat serves one zone of a "thermal load." Most large buildings have multiple zone systems, with air supplied to each zone specifically addressing the need and thermal load of the zone. For comparison on a smaller scale, a two-story home may have two zones, one for the lower floor and one for the upper floor, with one heating and cooling unit for each floor. The upper floor may have a higher thermal load and may require more cooling than the lower floor.

Components

HVAC systems can consist of a plethora of components and can be complex. The major components include chillers, air handling units (AHUs), air terminal units (ATUs), and variable air volume equipment (VAV).

Boilers

Boilers are used to heat air. However, because of the general increase of the efficiencies in HVAC systems, many systems simply "recover" wasted heat produced from the chiller or use smaller-scale versions of traditional boilers to generate heat.

COMPONENTS OF AN HVAC SYSTEM

Air handlers	Supply fans
Return fans	Return-relief fans
Relief fans	Outside air fans
Exhaust fans	Pumps
Hydronic boilers	Air/water cooled chillers
Fan coils	Expansion tanks
Air separators	Cooling towers
Heat exchangers	Cooling tower pumps
Dampers	Heating coils
Cooling coils	Filters
Plenums	Spaces
Roof hoods	Louvers
Terminal boxes	Panel radiation
Finned radiation	Unit heaters

Figure 28 Components of an HVAC System

Chillers

Chillers, or air conditioners, utilize heat exchanges and circulate fluid or gas to cool the air that is passed through it. Water chillers are popular in larger buildings because of their efficiency. Pumps supply and return the chilled or hot water. In the line returning to the chiller, a cooling tower draws heat out of the water. The pumps and chillers are often located in a mechanical area, at ground level, or in a central plant if within a campus environment.

Air Handling Units

Air Handling Units (AHUs) (Figure 29) provide warm or cool air to different parts of a building, using chilled water to cool the air or steam or hot water to heat the air. The AHU draws air into the unit, passes the air over heating and cooling coils, and then forces it through the air ducts. The AHUs have many of the networked points of the HVAC control system, managing air flow, heating, cooling and filtering. They can serve a building, a single floor on a building, or multiple floors of a building. If the AHU is serving multiple zones, each zone typically gets local control by having its own air premixed at the AHU.

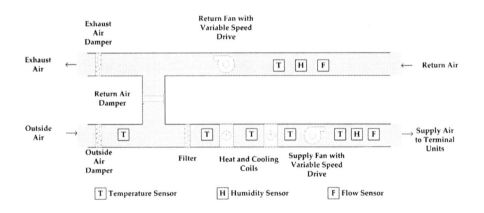

Figure 29 Typical Air Handling Unit

Air Terminal Units

Air Terminal Units (ATUs) (Figure 30) address specific HVAC thermal loads or zones. Thermal loads in a space can consist of exterior loads (outside air temperatures increasing or decreasing) and interior loads (people, lighting, computers, and other sources). ATUs compensate for these loads by providing air to the space at a specific temperature or quantity. ATUs make up for varying thermal loads by varying the temperature, varying the air volume, or doing both. VAVs can be pressure-independent, where the flow is maintained constant regardless of the inlet pressure, or they can be pressure dependent, where the flow rate of the VAV is dependent on the inlet pressure and, typically, the position of its damper or speed of a fan.

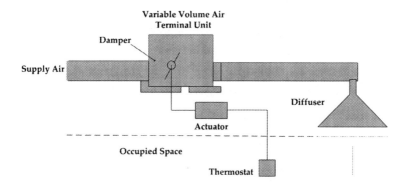

Figure 30 Variable Volume Air Terminal Unit

HVAC Controls

HVAC systems must control variable conditions of the system and its components. These conditions include liquid and gas pressure, temperature, humidity, the rate of flow of liquids and gases, and the speed and on/off state of mechanical equipment.

The HVAC system has a number of instruments and terminal devices available in the field that are used to gather data on the system and assist in controlling the system. The system controllers use input and data from sensor devices to make decisions about the system, and then, based on the input information, control actuator devices.

Sensors and transmitters include thermostats, liquid differential pressure transmitters for pumps and chillers, differential pressure sensors for fluids and airflow, static pressure sensors, air pressure sensors and humidity sensors. An example of an actuator or operator is an actuator for a damper that is mounted to the damper shaft and triggers the start of the damper operation. That operation could be a temperature sensor detecting a high temperature and sending a signal to the controller, which results in the controller sending a signal to an actuator to engage a motor that opens or closes a damper or vent.

These devices may communicate to each other or to the controller with analog or digital signals. Analog inputs to a controller can be a continuously changing signal from an external device or sensor, such as a temperature sensor. Digital inputs to a controller are simply a two-state, on-off signal from external devices or sensors, such as a switch.

In much the same way, analog outputs from a controller are "proportional variable" signals sent by the controller to adjust an actuator or external control device, such as a valve actuator. Digital output from a controller is a two-state or two-position signal from the controller to an actuator, such as control fan relay start-stop switch.

The field devices and equipment of building automation systems communicate at low network speeds, transmitting at rates of less than 1 Mbps. The communications network for a facility automation system is typically in a physical star or bus topology from the controller.

Some HVAC system controls can be provided through electric power or pneumatic means, usually with set or specific functions and a sequence of operation. However, direct digital control (DDC) is commonly used in more complex HVAC systems. DDC allows for a system controller to compute the sequence of operations based on the digital input from system sensors. Although DDCs are digital controls, they are able to handle analog-to-digital and digital-to-analog conversions. Unlike electric or pneumatic controls, DDC can be programmed for any sequence of operation.

Controllers can, confusingly, be referred to in a variety of ways: master, slave, terminal, floor, and others. The HVAC system network architecture typically consists of several network levels:

- Management Level
- System-Level or Building level Controllers
- Field-Level Controllers

Management Level

The top level of an HVAC control system is the management level consisting of personal computers or multiple PCs connected via an Ethernet network. These operator workstations can communicate with, interrogate, and control any of the controllers and devices on the network. The management level provides many functions:

- Administration and control of the HVAC system
- Programming for the system and other controllers, including sequences of operation
- Display of system information
- System reports
- System scheduling
- Archive and analysis of historical data
- Backup of controller databases
- Alarm reporting and analysis
- Trend analysis

The HVAC system is usually managed by a server and operator workstation using standard operating systems, specific HVAC software applications, GUI interfaces, and web access. The HVAC control system may be interfaced or integrated with fire alarm, video surveillance, access control, and lighting control systems. The HVAC system is also a significant part of a facility management and maintenance management system, primarily for tracking, managing, and optimizing energy use.

System-Level or Building Level Controllers

The system level or building level controllers are networked back to the management level. In a campus environment, the building level controllers are networked via a campus network to the management level of the HVAC control system. These controllers can manage HVAC equipment directly (typically, major components such as air handling units) or indirectly through networked downstream, lower-level controllers. System level controllers handle the operations of all downstream field level controllers, collect and maintain data, and can operate as stand alone units if communication is lost to the management level. System controllers have a peer-to-peer relationship with other controllers.

Field-Level Controllers

Field-level controllers serve building floors, and specific areas, applications, and devices. Field level controllers are limited controllers, in terms of both functionality and connectivity. Included in this group are DDC controllers, mechanical controllers, and application specific controllers.

DDC controllers may support multiple applications, specific device networks, or a particular equipment component, such as an air handling unit. The DDC controller usually has onboard memory, an operating system, and a database. Both DDC and mechanical controllers perform control through control algorithms. For example, the controller may measure temperature or humidity in a specific area and, based on the measurement, direct cooling, heating, humidifying, or dehumidifying to that area.

Some mechanical equipment, such as air handling units or chillers, may be procured with a field level controller and devices as part of the equipment. System controllers allow field controllers to communicate to other field controllers or a group of field controllers, and to access databases and programs. DDC controllers can also use remote application specific controllers (ASCs) for devices such as VAV terminal units.

HVAC control systems are evolving to the utilization of smart building infrastructure. This is true of the management level of the HVAC system and is evolving or trickling down to the network hierarchy at the system and field level. The adoption of ANSI/TIA/EIA 862 addressing structured cable infrastructure for building automation systems allows for standard unshielded twisted pair copper cabling and fiber optic cabling to be utilized throughout an HVAC control system. While the IP network protocol may be used at the management level, it competes primarily with BACnet and LonTalk at other levels. However, these other protocols recognize the dominance of the IP protocol and are either providing routers for transitioning from their native protocol to IP (LonTalk) or are migrating to IP with standards such as BACnet/IP.

Electric Power Management Systems

A facility's electric power management system (EPMS) monitors the power distribution system for usage and quality. The EPMS, together with the HVAC system and the lighting control system, are integral to the overall energy management of a facility to control usage and costs. In addition, the EPMS is a tool managing and ensuring the quality of the power, that is, a source of power that is free from surges, sags, and outages that may affect the reliability and safety of a facility.

The EPMS monitors the electrical distribution system, typically providing data on overall and specific power consumption, the quality of the power and event alarms. Based on that data, the system can assist in defining, and even initiating, schemes to reduce power consumption and power costs. The schemes "shed" power and are triggered by pre-determined thresholds, such as certain levels of power demand or a particular time-of day when resource consumption is high. The EPMS can manage event alarms, calculate usage trends, track and schedule maintenance, troubleshoot, and "bill back" metered power usage to specific users or tenants.

Typical systems monitor the power service entrance of a building or campus, switchgear, generators, network protectors, switchboards, panelboards, uninterruptible power supplies (UPS), emergency power generation, and more. The components of an EPMS include monitoring devices and control devices.

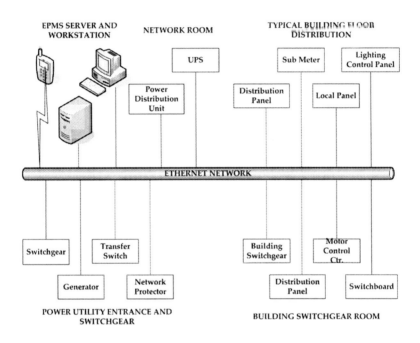

Figure 31 Electric Power Management System

Monitoring

The EPMS monitors the electric loads of major equipment. Inputs to the monitoring unit can be current or potential transformers as well as other sensors or monitoring devices. Current transformers are used to provide information on electric current, while potential transformers are used to provide information for electric voltage meters.

The monitoring units are microprocessor based, have onboard memory, and can be programmed or may have pre-set factors for monitoring, testing, and reporting sets. Some monitoring units can display locally and/or communicate via the EPMS network.

The monitoring unit may have inputs for specific distribution components, such as electrical breakers. Addressable relays may be part of a monitoring portfolio, to be used for sensing and the monitoring of devices where information is simply in an open/closed or on/off state. The monitoring of critical equipment, such as switchboards and switchgear, may include digital metering and the capability to monitor power quality. For power quality, most EPMS implementations adhere to the IEEE categories for power quality problems, including sags, swells, harmonics, interruptions, under and over voltages, and transients.

Display Units

Display units typically connect to the monitor of an electrical load or equipment. The displays can be local, specific to particular equipment, or allow for viewing and monitoring of multiple loads and devices. Some larger display units may connect to multiple monitors and be able to communicate to an operator's workstation via a standard data network.

Central Operator Workstation

A central operator workstation for the EPMS is usually a personal computer with special application software. It uses data from the system components to analyze and take action regarding the usage of power in a facility. The operator workstation can have the following features:

- Distribution system graphics
- Real time reports
- Trend reports
- Historical reports
- Alarm reporting
- Analysis of electric power waveforms
- Determination and initiation of power load-shedding strategies
- Communication with HVAC and lighting control systems
- Usage Billing Software

EPMS implementations have evolved to the infrastructure foundations of smart buildings. Backbone EPMS network connections can use standardized unshielded twisted pair copper and fiber optic cables, supplemented with shielded twisted pair copper cable. Physical topologies can be bus, star, and daisy chain configurations. Many of the lower level network communications utilize RS-485 communications, while others use TCP/IP protocols. EPMS network components and devices connect to the Ethernet network directly or through connectivity to a network router or gateway, essentially converting or encapsulating other network protocols, such as Modbus®. Databases for EPMS environments typically adhere to SQL standards. Systems also allow for connectivity to remote client personal computers and PDAs. The EPMS is an important facility operational tool and a critical element to a smart building.

Lighting Control Systems

Facility lighting is needed to provide visibility for building occupants, aesthetic atmosphere for spaces and rooms, and for life safety. It is estimated that lighting accounts for 30-40% of electricity usage and costs in a typical building. Therefore, unneeded and uncontrolled lighting within a building not only wastes energy but also increases facility operational costs. In addition, lighting can also affect other technology systems, such as the need and costs of cooling spaces where lighting has increased the space's temperature. Lighting control systems provide lighting for occupants of the building as needed, but do it in an efficient manner, consistent with any applicable building and energy codes.

The need for lighting in a building varies by the type of building, spaces within the building, time of day, and occupancy of the building. Consequently, the control strategies and functions of a lighting control system reflect these variances and primarily involve:

- Scheduling
  A control system may have a pre-determined schedule of when lights are turned on and turned off

113

- Occupancy Sensors
  For spaces of a building where occupancy is difficult to predict (such as meeting rooms or restrooms), lights may be turned on and off based on a lighting control system device sensing occupancy

- Daylight
  To reduce the need and cost of lighting spaces, a control system utilizes natural light as much as possible. This is sometimes called "daylight harvesting" or "daylight-ing."

The lighting control system distributes power to the available lighting units in a typical fashion, but inserts digital control and intelligence in many, if not all, of the devices controlling the lighting, such as the circuit breaker panel, wall switches, photocells, occupancy sensors, back up power and lighting fixtures. The control system significantly increases the functionality and flexibility of the lighting system by providing digital control and intelligence to the end devices. For example, a reconfiguration of lighting zones is accomplished through software rather than the physical re-cabling of the lighting zones. In additional, the intelligent end devices allow more focused application of lighting control needs and strategies to specific spaces within the building.

System Control

The heart of the lighting control center is typically a server, web-enabled and interconnected to other facility technology systems, and a workstation, with a GUI interface and client software for system administration. The networked system allows any authorized individual, including tenants or other occupants, to adjust their lighting through the network or a web browser.

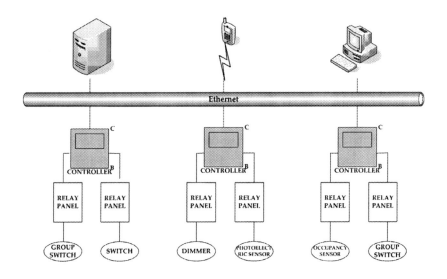

Figure 32 Lighting Control System

115

One approach to the lighting control system is the use of intelligent controllers. These controllers are distributed throughout a facility, house system electronics, and manage downstream relay panels. The controllers and the system server are networked via an Ethernet network (Figure 32), usually sharing schedules and overrides. The controller may have a user interface panel, which can be used instead of a system workstation, to program and monitor the lighting control system.

System controllers may be modular to allow for growth. The controller may also have several communications interfaces such as an Ethernet port, and ports for RS-232 and RS-485 communications. The system controller communicates with each of the panels through an Ethernet connection, or a BACnet or LonTalk protocol that is routed to a larger IP network.

Another emerging networking approach for lighting control systems is distribution of the intelligence and control further downstream to each device by providing a network interface for each device. This approach centralizes the control to the network server and allows for network interfaces to specific devices.

Relay Panels

Relay panels are typically mounted next to the electrical circuit breaker panels. The circuit breaker panel feeds into the relay panel, with the relays within the relay panel acting as a switching device for the circuit. Many relay panels can be fed by both 120V and 277V circuit breaker panels, and relay groups can be fed by different voltages within the same panel. Each relay can be individually programmed through the system controller or the relay panel.

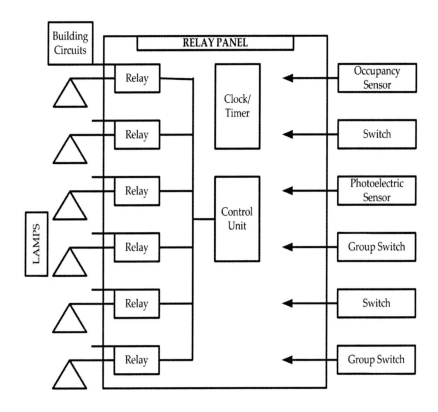

Figure 33 Lighting Control Relay Panel

The relay panels provide line voltage control of the lighting loads. Relay panels allow for a single circuit to feed into several relays, and allow multiple circuit breaker panels to feed into a single relay panel. While relay panels can be programmed or controlled by a system controller, they can also operate without the system controller. The relay panels typically have status indicators for the relay outputs, dry contact inputs for program override purposes, and inputs for monitoring devices, such as photocells and occupancy sensors.

In a multi-story facility, there may be a relay panel on each floor controlling all the lights on the floor. Each room on the floor has a local switch and there is also a master switch for the whole floor. The master switch for the floor may be programmed to turn lights on at 7AM and off at 6PM; between 6PM and 7AM, the system may repeatedly perform an "Off" sweep to turn lights out where the programming has been overridden.

Occupancy Sensors

Occupancy or motion sensors are devices that sense the presence or absence of people within their monitoring range. They may be used in restrooms, utility rooms, conference rooms, coffee rooms, locker rooms and many other spaces. Typically, the sensor and a control unit can be enclosed in one unit, such as a wallbox, but for larger facilities, the sensor is tied to a relay panel. The control unit or the relay is programmed to turn lights "on" when the presence of people is sensed by the motion detector, and may be programmed to turn the lights "off" if the space is unoccupied for a pre-determined time period. The sensitivity of the sensor may also be adjustable.

There are several types of motion sensors available, including passive infrared (PIR), active ultrasound, and hybrid technologies, such as PIR and active ultrasound, or PIR and audible sound. These sensors are typically used in locations such as hallways, lobbies, private offices, conference rooms, restrooms, and storage areas.

Ultrasonic sensors emit ultrasonic waves, sense the frequency of the reflected waves, and sense motion if there are shifts in the frequency of the reflected waves. Ultrasonic motion detectors provide continuous coverage of an area.

PIR sensors detect radiation, that is, the heat energy that is released by bodies. PIRs operate in a line of sight and have to "see' an area, so they can not be obstructed by open area partitions or tall furniture. PIRs use a lens to focus heat energy so that it may be detected. However, the lens views the covered area through multiple beams or cones and may create coverage gaps.

Other technologies and approaches to motion detection include sensing audible noise. Hybrid sensors (PIR and ultrasound, PIR and audible) offer the most effective occupancy detection and have maximum sensitivity without triggering false detections.

Dimmers

Dimmer modules manage low voltage switch and line voltage output controls of the dimmer's lighting loads. Stand-alone dimmers typically have status indicators, analog inputs for photocell or occupancy sensors, diagnostics, and are able to optimize responses for different types of lighting fixtures. Dimmers can be used for specific spaces, such as areas with audio-visual presentations, or throughout the total system for managing large facilities.

Like occupancy sensors, dimmer switches connected to a relay panel. Preset dimming controls from a relay panel provide predetermined dimming for several channels or loads. Presets are tamper-proof, that is to say, they will not allow anyone except the approved and authorized lighting control personnel to override the presets.

Dimming can be use to implement several energy savings strategies. For example, lights can be dimmed when the demand for electricity exceeds a pre-determined level, possibly as part of an overall load shedding policy. Such reductions are typically unnoticeable for most users. Another example involves fluorescent lamps. The light output of fluorescent lamps decreases over the life of the lamp (the expected depreciated output may be used as an initial design factor). Dimming can be used with new fluorescent lamps to produce the desired light level and then gradually manage the lighting level over the life of the lamps, to produce both a constant level of light output as well as a longer lamp life.

Photoelectric Controls

Photoelectric controls are designed to strategically use day-light to reduce the need for artificial lighting. They may be located in perimeter offices, atriums, hallways, or in areas with skylights. Ambient light sensors measure natural and ambient light, and based on the amount of natural light, adjust the lighting to maintain a constant light level. In some spaces, manual or automatic blinds, or other means of reducing the direct solar exposure glare, excessive light levels, and heat gain, can be used to supplement the photoelectric control. These may include motorized window shades or blackout shutters.

BAS Systems

An override of the programmable lighting control system may be triggered by the fire alarm system or emergency power generators. In the case of a fire alarm or loss of normal power, the lighting control system may turn on key emergency lighting fixtures.

Programming lighting control systems use smart technology infrastructure, such as unshielded twisted pair cables, Ethernet, and TCP/IP protocols. Because of the significant savings on overall energy usage and costs, lighting control systems are integral to smart buildings.

Facility Management Systems

A Facility Management System (FMS) is an overarching management system of the smart building technology systems. It brings together some of the operational management functions of the systems on a facility-wide, rather than a system-wide, basis.

The FMS is typically a server-based configuration coupled with operator workstations, which can be desktop, laptop, or wireless devices. Access to an FMS can also be achieved through Internet web access. FMS environments operate on a standard Ethernet IP network using a structured cable infrastructure. Operating systems are industry standards and databases, and are SQL and ODBC compliant. Smart building systems that do not make use of IP protocols must route other protocols (such as LonTalk or BACnet) to the IP protocol for the FMS.

The definition of a facility management system is somewhat ambiguous and can be confusing, but a facility management system in the context of a smart building usually focuses on one of two objectives.

One type of facility management system focuses on the "business processes" of facility management, that is, a tool that assists in managing service orders, inventory, procurement, and assets.

These systems are generally vended by companies with specific products for one of the technology systems in a building or by companies that are involved in broader business processes products as well (human resources, finance, purchasing, etc.).

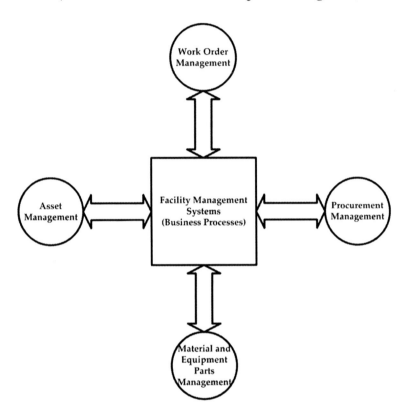

Figure 34 FMS - Business Processes

The other FMS type is more focused on the "operational functions" of the smart building systems, primarily life safety and building automation systems. The systems are typically vended by the manufacturer of the building automation and life safety systems. In many cases, facilities need both types and will deploy and integrate the systems as needed.

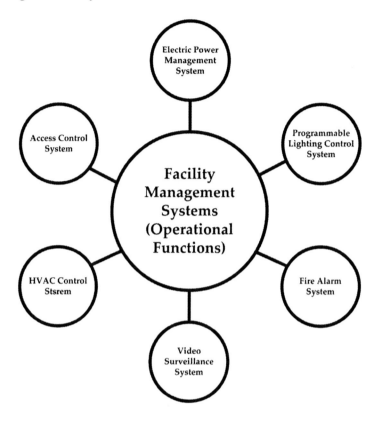

Figure 35 FMS Operation Processes

The "business process" facility management system is a series of software modules. These allow an owner or facility manager to select modules based on operational and facility management need. These modules may include:

- Work Order Management

A work order module tracks and monitors all service requests. A work order management application:

- o Tracks who an action is assigned to
- o Tracks what the current status of actions are
- o Keeps records of updates on the actions
- o Tracks when the action is completed
- o Keeps an archive of the work that can be referenced later

Through the use of wireless devices, a system can notify technicians and other mobile employees with service requests or other work order data on any text-messaging or email device, such as cell phones, PDAs, tablet personal computers or pagers. Systems may allow for tenants or "customers" to send service requests over the network or Internet and be automatically notified on the status of the requests.

A work order system tracks labor, materials, and travel expenses for work orders, attaches documents to work orders, projects labor demands for any future time period based upon the work orders, and monitors work-performance indicators in addition to cost or time.

- Asset Management

An asset management module manages all necessary equipment data, including the name of equipment, serial numbers, location, vendor, internal cost center, warranties, equipment performance, and equipment documentation.

- Material and Equipment Parts Management

A material management system tracks material and equipment parts that are moved into, moved out of, or that currently reside in inventory. It ensures that the right level of inventory is maintained. The material management module is typically tied to and integrated with the work order system.

- Procurement Management

Procurement management automates and streamlines the procurement process for vendor services and equipment. Contractors are pre-approved for specific types of services and equipment. Then, for example, the work order management system, can automatically assign a work request to a pre-approved external contractor, triggering and tracking the procurement process needed. Further, the system can learn that needed materials and equipment are not in inventory, and issue a direct purchase from a contractor to fulfill an outstanding work order.

Procurement management has other benefits. Some systems can track contracts, key contract dates, and the performance of contractors. Systems can also automatically requisition materials and equipment based on pre-determined maintenance schedules.

An "operational processes" facility management system monitors, supervises, controls, and reports on other smart building systems. These systems may include access control, video surveillance, fire alarm, HVAC control, programmable lighting, and electric power management. Its basic functions are:

- Providing information on those supervised building functions including, but not limited to, current status, archived historical information, summaries, analysis, displays, and reports on control and management functions
- Detecting, annunciating, and managing alarm and other conditions
- Diagnostic monitoring and reporting of system functions, nodes, devices, and communication networks
- Interfacing between individual smart building applications.

These systems typically display the following responses to an operator's workstation:

- Alarm summary
- Event summary
- Trend set displays
- Group control and group trend displays
- Communications status
- System status
- Configuration displays
- Communication links status
- System parameters configuration
- Time schedule assignment
- Holiday assignment
- History assignment

- Events archive and retrieval
- Time period summary and configuration
- Point detail for every configured point

Energy Management System

An energy management system (EMS) is used to reduce a facility's energy and electric costs, while still maintaining a comfortable and safe environment for a building's occupants. As part of a smart building, the EMS brings together and addresses the main electric and energy systems, namely the HVAC system, the lighting control system, and the power management system.

Electrical utilities base their charges on several factors, but the most important are power consumption and demand. Consumption is simply the total amount of electricity used in a billing period. Demand is typically the cost per kilowatt depending on time or season. So, reducing consumption and managing demand are basic strategies of an energy management program. An energy management program coordinates the HVAC and lighting control systems, together with an equipment maintenance program, to achieve optimum energy usage.

The EMS may be a software module within a larger facility management system, or a separate stand alone application such as energy management being an integral part of the HVAC control system. An EMS usually provides a group of applications to optimize a facility's energy usage and costs. This may include the programs such as:

- Tracking of utility bills to monitor the usage and costs, as well as comparing projected, budgeted, and actual usage and costs
- Comparing the energy usage to other similar buildings, an exercise known as "benchmarking"
- Calculating "thermal comfort," the cost for different lighting control strategies, proper ventilation rates, and so forth.

Because energy can be saved by making sure equipment is operating in the most cost effective manner, an EMS may have a feature to ensure proper maintenance of mechanical and electrical equipment. Maintenance can be scheduled at regular intervals from historical data or based on the manufacturer's recommendations, or equipment can be monitored to determine abnormal operating conditions.

Monitoring requires setting acceptable operating ranges for equipment and then setting alarms if the equipment operates out of range. Equipment indicators that may signal possible malfunctions include:

- Temperature
- Vibration
- Pressure
- Air quality
- Humidity
- Energy consumption
- Gases

Maintenance Management Systems

Computerized Maintenance Management Systems (CMMS) are comprehensive software applications for a variety of equipment and materials. Like energy management systems, a CMMS implementation can be integrated with a software module of a larger facility management system or be a separate stand-alone application.

The CMMS integrates or communicates with the FMS system to retrieve data from field devices in determining when equipment or materials require maintenance. The CMMS triggers an automatically generated work order request for a piece of equipment when certain maintenance conditions are met. Types of configurable maintenance events may include duty cycles, run hours, and high data values.

A CMMS may include a fleet management module that can automatically create work orders for preventative maintenance of vehicles, establish a vehicle replacement schedule based upon vehicle cost, as well as track registration, license renewal, and warranty work.

Index

Appendix: Organizations and Associations

AEE - Association of Energy Engineers
4025 Pleasantdale Rd., Suite 420
Atlanta, Georgia 30340
AEE is a profession association involved in energy efficiency, utility deregulation, facility management, plant engineering, and environmental compliance, and also provides certification programs.
www.aeecenter.org

AFE - Association of Facility Engineering
8160 Corporate Park Drive, Suite 125
Cincinnati, Ohio 45242
AFE is a professional organization that provides education, certification, technical information for plant and facility engineering, operations and maintenance professionals
www.afe.org

AIA – The American Institute of Architects
1735 New York Ave., NW
Washington, DC 20006-5292

AIA is the professional organization for architects in the United States.

www.aia.org

AIIB - Asian Institute of Intelligent Buildings
Department of Building & Construction
City University of Hong Kong
83 Tat Chee Avenue
Kowloon Tong, Kowloon
Hong Kong

AIIB is an independent and academic institute whose aim is to develop Asia's definition and standards for intelligent buildings and to act as an independent certification authority for intelligent buildings.

www.aiib.net

ANSI - American National Standards Institute
1819 L Street, NW Suite 600
Washington, DC 20036 USA

ANSI is a private, non-profit organization that administers and coordinates the U.S. voluntary standardization and conformity assessment system.

www.ansi.org

ASHRAE – The American Society of Heating, Refrigerating and Air-Conditioning Engineers
1791 Tullie Circle, N.E.
Atlanta, GA 30329

ASHRAE is an organization providing research, standards, publications and education on heating, refrigerating and air-conditioning.

www.ashrae.org

ASIS - ASIS International (previously the American Society of Industrial Security)
1625 Prince Street
Alexandria, VA 22314-2818
ASIS is an international professional organization involved with all aspects of security and provides certification for security professionals.
www.asisonline.org

ASTM - American Society for Testing and Materials
100 Barr Harbor Drive
PO Box C700
West Conshohocken, PA, 19428-2959
ASTM International is one of the largest voluntary standards development organizations for technical standards for materials, products, systems, and services.
www.astm.org

BICSI - Building Industry Consultant Services International
8610 Hidden River Pkwy.
Tampa, FL 33637-1000
BICSI is an international telecommunications organization providing education professional registration programs.
www.bicsi.org

BOMA – Building Owners and Management Association
1201 New York Avenue, NW
Suite 300
Washington, DC 20005

BOMA is an international organization of building owners, managers, developers, facility managers and other professionals and provides education on office building development, leasing, building operating costs, energy consumption patterns, local and national building codes, legislation, occupancy statistics and technological developments.
www.boma.org

CABA – Continental Automated Building Association
1200 Montreal Road
Building M-20
Ottawa, ON
K1A 0R6
CABA is a not-for-profit industry association that promotes advanced technologies for the automation of homes and buildings in North America.
www.caba.org

CANSA – Canadian Security Association
610 Alden Road, Suite 100
Markham, Ontario, Canada L3R 9Z1
CANSA is a Canadian organization providing security education, government relations, marketing, communications, leading industry trade shows and the latest industry information and news.
www.canasa.org

CSA International – Canadian Standards Association
178 Rexdale Boulevard
Toronto, Ontario, CANADA, M9W 1R3

CSA tests and certifies products to Canadian, US and other nations' standards and issues the CSA Mark for qualified products.
www.csa-international.org

CSC - Construction Specifications Canada
120 Carlton Street, Suite 312
Toronto, ON, M5A 4K2
CSC is a national association dedicated to the improvement of communication, contract documentation, and technical information in the construction industry, providing publications, education, professional development, and certification.
www.csc-dcc.ca

CSI - The Construction Specifications Institute
99 Canal Center Plaza, Suite 300
Alexandria VA 22314
CSI is a professional organization with a mission to advance the process of creating and sustaining the built environment. CSI provides information and education, and also produces the MasterFormat, the National CAD Standards and the Project Resource Manual that are utilized for the design and construction of buildings.
www.csinet.org

EIA - Electronic Industries Alliance
2500 Wilson Blvd.
Arlington, VA 22201
EIA is an organization of electronic and high-tech associations and companies whose mission is promoting the market development and competitiveness of the U.S. high-tech industry through domestic and international policy efforts. EIA focuses on the areas

of innovation and global competitiveness, international trade and market access, telecom and information technology, and cyber security.

www.eia.org

ETSI – European Telecommunications Standards Institute
650, route des Lucioles
06921 Sophia-Antipolis Cedex
France
ETSI is an independent, non-profit organization responsible for standardization of information and communication technologies within Europe, including telecommunications, broadcasting and related areas such as intelligent transportation and medical electronics.

www.etsi.org

FCC - Federal Communications Commission
The FCC is a U.S. government agency responsible with regulating interstate and international communications.
445 12th Street, SW
Washington, DC 20544
www.fcc.gov

ICC - International Code Council
5203 Leesburg Pike, Suite 600
Falls Church, VA 22041-3401
ICC is a nonprofit organization dedicated to developing a single set of comprehensive and coordinated national model construction codes.

www.iccsafe.org

ICIA - InfoComm International Association
11242 Waples Mill Road, Ste. 200
Fairfax, VA 22030
InfoComm is the international association of the professional audio visual industries providing education and certification for the AV market.
www.infocomm.org

IEEE - Institute of Electrical and Electronic Engineers
445 Hoes Lane
Piscataway, NJ 08854-1331 USA
IEEE is a non-profit organization that promotes the engineering process of creating, developing, integrating, sharing, and applying knowledge about electro and information technologies and sciences. IEEE is a source of technical and professional information, and standards.
www.ieee.org

IES - Illuminating Engineering Society
IES of North America
120 Wall Street, Floor 17
New York, NY 10005
The IES has chapters throughout the world and is the recognized technical authority on illumination, providing information to designers, manufacturers, engineers and researchers.
www.iesna.org

IFMA - International Facility Management Association
1 E. Greenway Plaza, Suite 1100
Houston, TX 77046-0194

IFMA is a professional association for facility management, with members in 56 countries. IFMA certifies facility managers, conducts research, provides educational programs, and recognizes facility management degree and certificate programs.
www.ifma.org

ISA – Instrument Society of America
67 Alexander Drive
Research Triangle Park, NC 27709
ISA is a professional organization that develops standards, certifies industry professionals, provides education and training, focusing on instruments, systems and industrial automation.
www.isa.org

ISO - International Organization for Standardization
1, rue de Varembé, Case postale 56
CH-1211 Geneva 20, Switzerland
ISO is a non-governmental organization consisting of the national standards institutes of 156 countries, with a Central Secretariat in Geneva, Switzerland, that coordinates the system. ISO is the world's largest developer of technical standards.
www.iso.org

LCA - Lighting Control Association
1300 North 17th Street, Suite 1847
Rosslyn, VA 22209
The Lighting Controls Association (LCA) is an adjunct of the National Electrical Manufacturers Association (NEMA). LCA provides education for the professional building design, construction and management communities about the benefits and operation of automatic switching and dimming controls.
www.aboutlightingcontrols.org

NFPA - National Fire Protection Association
1 Batterymarch Park
Quincy, MA 02169-7471
NFPA is an international organization providing and advocating consensus codes and standards, research, training, and education. NFPA's codes and standards influence every building in the United States, as well as many of those used in other countries. NFPA's code development process is accredited by the American National Standards Institute (ANSI).
www.nfpa.prg

NSCA – National Systems Contractors Association
625 First Street SE - Suite 420
Cedar Rapids, IA 52401
NSCA is professional association representing the commercial electronic systems industry focusing on low-voltage systems.
www.nsca.org

SCTE - Society of Cable Telecommunications Engineers
140 Philips Rd.
Exton, Pa 19341-1318
SCTE is a professional association dedicated to cable television and telecommunications professionals. The organization provides professional development, information and standards. SCTE is accredited by the American National Standards Institute (ANSI).
www.scte.org

SFPE - Society of Fire Protection Engineers
7315 Wisconsin Avenue, Suite 620E
Bethesda, MD 20814

SFPE is a professional society representing those practicing the field of fire protection engineering.
www.sfpa.org

SIA - Security Industry Association
635 Slaters Lane, Suite 110
Alexandria, VA 22314-1108
SIA is an international association providing education, research, technical standards for the security marketplace.
www.siaonline.org

TIA - Telecommunications Industry Association
500 Wilson Blvd., Suite 300
Arlington, VA 22201-3834
TIA represents providers of communications and information technology products and services and provides standards development and advocacy with governments. The TIA is ANSI accredited.
www.tiaonline.org

UL - Underwriters Laboratories
333 Pfingsten Road
Northbrook, IL 60062-2096 USA
UL provides product-safety testing and certification within the United States.
www.ul.org

Printed in the United States
69526LVS00006B/98

9 780978 614409